Cor Ouwerkerk

Theory of Macroscopic Systems

A Unified Approach for Engineers, Chemists and Physicists

With several Figures

Springer-Verlag Berlin Heidelberg New York
London Paris Tokyo Hong Kong Barcelona

Cor Ouwerkerk

Northgodreef 15
NL-2202 XZ Noordwijk

ISBN-13: 978-3-540-51575-3 e-ISBN-13: 978-3-642-75010-6
DOI: 10.1007/978-3-642-75010-6

Library of Congress Cataloging-in-Publication Data
Ouwerkerk, Cor. Theory of macroscopic systems : physical, chemical and engineering systems / Cor Ouwerkerk. p. cm. Includes bibliographical references.
ISBN-13: 978-3-540-51575-3 (U.S.)
1. Engineering – Mathematics. 2. Thermodynamics – Mathematics. 3. Fluid mechanics – Mathematics. 4. Chemical engineering – Mathematics. I. Title. II. Title: Macroscopic systems.

Typesetting: Macmillan India Ltd., Bangalore;

2131/3020-543210 – Printed on acid-free paper

Preface

This book deals with the theory of macroscopic systems. Traditionally this theory has been fragmented over a number of disciplines like thermodynamics, physical transport phenomena, sometimes referred to as non-equilibrium or irreversible thermodynamics, fluid mechanics, chemical reaction engineering and heat and power engineering. This fragmentation, the different approaches followed in presenting theory, e.g. the inductive approach as opposed to the postulational approach in textbooks on thermodynamics, many alternative representations of equations and differences in notation make it cumbersome to discern a single coherent theory of macroscopic systems.

The idea of this book is to present the theory of macroscopic systems as a unified theory with equations strictly developed from a single set of principles and concepts.
The book is an attempt to bridge gaps between the various disciplines. It can serve as a textbook, refresher or reference book to students of an advanced level in various disciplines, to scientists and to practising engineers working in design and development. It provides rigorous equations and their possible simplifications for use in computer models for scale-up or optimisation. Topics like exergy analysis and multi-component diffusion are included.

The principles and concepts in the theory of macroscopic systems comprise in addition to the mole and mass balances over a system, the balance equations for the fundamental extensive properties momentum, energy and entropy as well as the phenomenological laws on asymptotic phase behaviour and molecular transport.
The fundamental extensives are defined by their transport through a boundary other than by matter, and by their production in a system. These definitions include forces, work, heat and temperature. The production of entropy in a system is postulated to be zero when the system is at equilibrium or passes reversibly through a series of equilibrium states. It is positive in all other cases.
The laws on molecular transport are postulated on the basis of the volumetric rate of entropy production. We restrict ourselves to linear transport laws to obtain Newton's law for momentum and the Standart–Taylor–Krishna equations for matter and heat.

In the *list of symbols* nomenclature is adopted to reflect coherence of symbols and equations by their appearance, to minimise the number of subscripts and superscripts and, finally, to coincide as much as possible with existing and recommended nomenclature.

Structure and contents gives the pattern by which equations are derived from the principles of the theory of macroscopic systems as well as a broad summary of the book. The equations formulated and derived for a variety of systems include balance equations, property relations and equilibrium conditions, equations of property change and equations for transport and production.

The individual equations formulated and derived are not original. References often consulted in writing the book are summarised at the end of this preface. With some of the authors I discussed a few theoretical imperfections in their books. Usually they are of minor practical consequence.

I wish to express my thanks to a number of people who encouraged me in the early stages of the development of the book, were just patient enough to listen, or commented on the almost final manuscript. They include dr.ir.G. Ooms of the Koninklijke/Shell Laboratorium Amsterdam and Professors J. Villermaux of the Laboratoire des Sciences du Genie Chimique in Nancy, J. de Swaan Arons and J.A. Wesselingh of the Technical University Delft, J.H.M. Fortuin of the University of Amsterdam, and W.P.M. van Swaaij of the Technical University Twente. My thanks go as well to Mrs. Hetty de Bruijn-Janssen of the Koninlijke/Shell Laboratorium Amsterdam for the skilful typing of the manuscript and her patience. My wife Aagje deserves thanks for allowing me to steal the time to develop and write this book.

Noordwijk, October 1990 Cor Ouwerkerk

References

Thermodynamics

E.A. Guggenheim, "Thermodynamics", North-Holland Publishing Company, Amsterdam, 1950.
F.E.C. Scheffer, "Heterogene evenwichten in unaire en binaire stelsels", Uitgeverij Waltman, Delft.
K. Denbigh, "The principles of chemical equilibrium", Cambridge University Press, 1971.
J.M. Smith, H.C. Van Ness, "Introduction to chemical engineering thermodynamics", McGraw-Hill Kogakusha., Tokyo, 1975.

Physical Transport Phenomena

R.B. Bird, W.E. Stewart, E.N. Lightfoot, "Transport phenomena", John Wiley, New York 1963.
G.L. Standart, R. Taylor and R. Krishna, "The Maxwell–Stefan formulation of, irreversible thermodynamics for simultaneous heat and mass transfer", Chem. Eng. Commun. Vol. 3, pp. 277–289, 1979.
W.J. Beek, K.M.K. Muttzall, "Transport phenomena", John Wiley, Chichester, 1980.

Chemical Reaction Engineering

K.R. Westerterp, W.P.M. van Swaaij and A.A.C.M. Beenackers, "Chemical reactor design and operation", John Wiley, Chichester, 1984.

Heat and Power Engineering

C. Osborn Mackay, W.N. Barnard, F.O. Ellenwood, "Engineering thermodynamics", John Wiley, New York, 1957.

Contents

List of Symbols

Generalised Extensive Properties

E	extensive property, [E]
ΔE_M	extensive property of mixing (relative to pure components), [E]
E^{is}	extensive ideal-state property, [E]
E^{ig}	extensive ideal-gas property, [E]
E^{∞}	extensive infinite-dilution state property, [E]
E^{z}	extensive zero-dilution state property, [E]
E^{ni}	extensive non-ideality property, [E]
E^{r}	extensive residual property (relative to ideal-gas state), [E]
$E^{e,\infty}$	extensive excess property relative to infinite-dilution state, [E]
$E^{e,z}$	extensive excess property relative to zero-dilution state, [E]

Molar, Specific and Volumetric Properties of a Generalised Extensive E

$\bar{E}$	average molar single-phase property, $[E].kmole^{-1}$
$\bar{E}_i$	pure-component molar single-phase property, $[E].kmole^{-1}$
$\bar{e}$	average specific single-phase property, $[E].kg^{-1}$
$\check{E}$	concentration of E, $[E].m^{-3}$
$\tilde{E}$	average molar multi-phase property, $[E].kmole^{-1}$
$\tilde{e}$	average specific multi-phase property, $[E].kg^{-1}$
$\Delta\bar{E}$	property of phase transition, $[E].kmole^{-1}$
$\Delta\bar{E}$	$= \sum \nu_i E_i$, property of reaction, $[E].kmole^{-1}$

Partial and Standard Properties of a Generalised Extensive E

E_i	partial molar property, $[E].kmole^{-1}$
e_i	partial specific property, $[E].kg^{-1}$
$\check{E}_i$	partial concentration, $[E].m^{-3}$
E_i^0	standard property, $[E].kmole^{-1}$
ΔE_i^0	standard property of phase transition, $[E].kmole^{-1}$
$\Delta E_{f,i}^0$	standard property of formation, $[E].kmole^{-1}$
ΔE^0	standard property of reaction, $[E].kmole^{-1}$

Transport and Production of a Generalised Extensive E

$\underline{\dot{E}}''_x$	flux of total transport (exchange) of E, $[E].s^{-1}.m^{-2}$
$\dot{E}_x$	rate of total transport of E, $[E].s^{-1}$
δE_x	total transport of E in differential time interval dt, [E]
E_x	total transport of E in a finite time interval, [E]

$\underline{\dot{E}}''$, $\dot{E}$, δE:	transport of E by matter
$\underline{\dot{E}}''_{im}$, $\dot{E}_{im}$, δE_{im}, E_{im}:	immaterial transport of E (transport of E other than by matter)
$\dot{E}'''_p$	volumetric rate of production of E, $[E].s^{-1}.m^{-3}$
$\dot{E}_p$	rate of production of E, $[E].s^{-1}$
δE_p	production of E in differential time interval dt, [E]
E_p	production of E in a finite time interval

Symbols

a	thermal diffusivity, $m^2.s^{-1}$
a	number of additional conditions
a_W	volumetric wall area of a flow channel, $m^2.m^{-3}$
A	surface area, m^2
A_i	constant in $C^0_{p,i} = A_i + B_iT$
A_i	diffusion potential, $J.kmole^{-1}$
$\underline{\nabla} A_i$	gradient of diffusion potential, $N.kmole^{-1}$
$\underline{\nabla} A_{ij}$	gradient of potential for diffusion of i relative to j, $N.kmole^{-1}$
b_{li}	coefficient of element l in equation of formation for one mole of component i
B_i	constant in $C^0_{p,i} = A_i + B_iT$, K^{-1}
BE	balance equation of a system for an extensive property
c	total molar concentration, $kmole.m^{-3}$
c_i	component molar concentration, $kmole.m^{-3}$
C_p	heat capacity at constant pressure, $J.K^{-1}$
C_v	heat capacity at constant volume, $J.K^{-1}$
D_{ij}	Maxwell–Stefan diffusivity, $m^2.s^{-1}$
$\mathbb{D}_i$	Fick's law diffusivity, $m^2.s^{-1}$
D/Dt	substantial derivative operator
$\mathscr{D}$	diffusivity for momentum, matter or heat, $m^2.s^{-1}$
e	$= e_1 - e_2$, equilibrium electrochemical potential, V
e_i	electrode potential, V
f	Fanning friction factor
f	number of independent properties or degrees of freedom
f_{ext}	minimum number of extensive independents or degrees of freedom
f_{int}	number of intensive independents or degrees of freedom
$\underline{f}_i$	partial specific force, $N.kg^{-1}$
F	free energy, J

$\underline{F}$	force or rate of immaterial transport of momentum, N or $kg.m.s^{-2}$
$\underline{F}_i$	partial molar force, $N.kmole^{-1}$
$\check{\underline{F}}_i$	force on component i per unit volume or volumetric rate of production of component momentum, $N.m^{-3}$ or $kg.m^{-2}.s^{-2}$
FBE	fundamental balance equation
FPR	fundamental property relation
$\mathscr{F}$	Faraday's constant, 0.965×10^8 C.(kg equivalent)$^{-1}$
g	gravitational acceleration, $N.kg^{-1}$ or $m.s^{-2}$
G	free enthalpy, J
h	height, m
H	enthalpy, J
$\mathscr{H}$	Henry's law constant, $N.m^{-2}$
k	$= R/N_{av}$, Boltzmann's constant, $J.molecule^{-1}.K^{-1}$
k_i	mass transfer coefficient for component i, $m.s^{-1}$
K	standard chemical equilibrium constant corrected for effect of pressure on chemical potentials of condensed phases
K^0	standard chemical equilibrium constant
K_T	isothermal compressibility, $N^{-1}.m^2$
K_S	isentropic compressibility, $N^{-1}.m^2$
l_i	molality, kmole.(kg solvent)$^{-1}$
L	length, m
m	total mass, kg
m_i	component mass, kg
$\dot{\underline{m}}''_{iD}$	diffusion flux of component mass relative to mass average velocity, $kg.s^{-1}.m^{-2}$
$\bar{M}$	average molar mass, $kg.kmole^{-1}$
M_i	component molar mass, $kg.kmole^{-1}$
$m\underline{v}$	momentum, $kg.m.s^{-1}$
n	total mole number, kmole
n_i	component mole number, kmole
$\dot{\underline{n}}''_{iD}$	diffusion flux of component mole number relative to mole average velocity, $kmole.s^{-1}.m^{-2}$
n_{kp}	production of n_k, the mole number of component *k*, in a finite time interval, kmole
N	number of components
N′	apparent number of components
N_{av}	Avogadro's number, 6.02×10^{26} molecules.kmole^{-1}
Ox_i	oxidant of redox system
p	pressure, $N.m^{-2}$

$\bar{p}_i$	equilibrium pressure of single-component two-phase system, $N.m^{-2}$
PR	property relation of a system
q	electrical charge transported in a finite time interval, C
Q	heat into a system in a finite time interval, J
Q^-	heat from a system in a finite time interval, J
r	number of independent reactions
r	radius, m
R	gas constant, 8.3143 $kJ.kmole^{-1}.K^{-1}$
Red_i	reductant of redox system
S	entropy, $J.K^{-1}$
t	time, s
T	temperature, K
$\bar{T}_i$	equilibrium temperature of single-component two-phase system, K
U	internal energy, J
$\underline{v}^*$	mole average velocity, $m.s^{-1}$
$\underline{v}$	mass average velocity, $m.s^{-1}$
$\underline{\bar{V}}''$	volume average velocity, $m.s^{-1}$
$\bar{V}_x$	volume exchange (displacement), m^3
V	volume, m^3
W	work on a system in a finite time interval, J
W^-	work by a system in a finite time interval, J
$\underline{x}$	place vector, m
x_i	mole fraction
X	$= mgh + \frac{1}{2}mv^2$, mechanical energy, J
y_i	mole fraction in gas phase
z	coordinate in flow direction, m
z_i	number of units of electrical charge of an ion
Z	$= pV/(RT)$, extensive compressibility factor, kmole
$\bar{Z}$	$= p\bar{V}/(RT)$, compressibility factor
α	thermal expansion coefficient, K^{-1}
α	heat transfer coefficient, $W.m^{-2}.K^{-1}$
α_{ij}	thermal diffusion factor for component pair $i-j$, m^{-1}
γ_i	activity coefficient
$\underline{\underline{\delta}}$	unit tensor
$\underline{\delta}_1$	unit vector
$\underline{\delta}_1\underline{\delta}_m$	unit dyad
ε	exergy, J

κ	$= C_p/C_v$, ratio of heat capacities at constant pressure and constant volume
λ	thermal conductivity, $W.m^{-1}.K^{-1}$
μ	dynamic viscosity, $N.m^{-2}.s$
μ_i	chemical potential, $J.kmole^{-1}$
ν	kinematic viscosity, $m^2.s^{-1}$
ν_i	stoichiometric coefficient relative to a key reaction component
ν^α	phase-transition coefficient
ρ	total mass concentration or density, $kg.m^{-3}$
ρ_i	component mass concentration or density, $kg.m^{-3}$
τ	residence time, s
$\underline{\underline{\tau}}$	momentum-flux by shear or stress tensor, $N.m^{-2}$
$\underline{\tau}_1$	momentum flux by shear in l-direction or stress on plane perpendicular to l-direction, $N.m^{-2}$
τ_{lm}	m-component of $\underline{\tau}_1$, $N.m^{-2}$
φ	angle
φ	expansion ratio of piston-in-cylinder system
φ_i	fugacity coefficient
$\bar{\varphi}_i$	pure-component fugacity coefficient
ϕ	electrical potential, V
ω	number of coexistent phases
ω	angular velocity, s^{-1}
$\underline{\nabla}$	vector differential operator (nabla)

Dimensionless Groups

Fo	$= \mathscr{D}t/L^2$, Fourier number
Nu	$= \alpha L/\lambda$, Nusselt number
Pr	$= \nu/a$, Prandtl number
Re	$= \nu_\infty L/\nu$, Reynolds number
Sc	$= \nu/\mathbb{D}_i$, Schmidt number
Sh	$= k_i L/\mathbb{D}_i$, Sherwood number

Subscripts

i, j, k	components
l, m, n	coordinate indices
cp	condensed phase
D	diffusion
f	formation

fr	friction
im	immaterial
M	mixing
p	production
w	wall
x	exchange
∞	far away from wall or interface

Superscripts

e	excess
c	concentration based
G	gas phase
ig	ideal gas
is	ideal state
l	molality based
L	liquid phase
ni	non-ideality
r	residual
S	solid phase
z	zero-dilution
α, ω	phases
0	standard
*	zero-exergy environment of a system
∞	infinite dilution

Structure and Contents

The accompanying table at the end of the volume summarises the structure and contents of the book. It has nine rows representing the chapters and three columns in which principles and concepts are consecutively postulated or applied. The first two of the nine chapters deal with principles and first deductions, the next three with property relations (PR's) of systems at equilibrium and the last four with balance equations (BE's) of systems distinct by the way their properties change with time and place. Column 1 involves mole numbers, masses and generalised extensive properties, column 2 the fundamental extensives momentum, energy and entropy and column 3 the phenomenological laws on asymptotic phase behaviour and molecular transport.

In the sections of the cells in the table equations are postulated or derived. In doing so input comes in general from cells with lower or equal row and column numbers, while text is presented on left hand pages and mathematical equations on right hand pages. Each chapter is preceded by a summary which is sometimes divided into summaries of the cells in the chapter. An appendix on vectors and tensors is included.

In *Chapter 1* the principles and concepts are introduced. For a generalised extensive property E of a system BE's, PR's, associated intensive properties and transport by matter are formulated. The fundamental extensives are defined by their immaterial transport (other than by matter) through a boundary and their production in a system. This involves forces, work, heat and temperature. The entropy production in a system is zero when it is at equilibrium or passes reversibly through equilibrium states. Otherwise it is always positive.

In cell 1.3 the infinite-dilution and the ideal-gas laws are postulated as PR's for chemical potentials.
The laws on molecular transport of momentum, matter and heat complete the column-3 principles. They are postulated on the basis of the column-2 equation for the volumetric rate of entropy production derived in Sect. 9.2.3 which reflects that each transport flux is down the gradient of its potential. The linear transport laws postulated are Newton's law for momentum and the Standart–Taylor–Krishna equations for matter and heat. They are only applied in cell 9.3.

In *Chapter 2* the BE's of a system are further formulated. For momentum, energy and entropy they are referred to as the fundamental BE's or FBE's. Application of the BE's to suitably devised systems and equating

the calculated entropy production to zero yield equilibrium conditions and PR's involving chemical potentials. For intraphase physical equilibrium the fundamental PR (FPR) of matter presents itself.
The exergy balance provides a tool particularly suited for *exergy analysis* of continuous flow processes. The exergy loss in a system or one of its subsystems represents and locates the loss of potential work due to irreversibly proceeding processes. The exergy analysis of a system includes a suitably selected environment with given temperature, pressure and components in abundant supply of zero exergy. PR's needed to calculate the partial exergy of a component in a stream for a given environment from tabulated standard property values are presented in Sect. 3.3.4.

Chapters 3, 4 and 5 are in the domain of *thermodynamics*. The PR's derived connect functions of state, express functions of state in terms of measurable properties or relate measurables.
The *set of single-phase PR's* in Sect. 3.2.3 summarises the numerous alternative representations of the FPR and PR's of the form $dE(T, p, n_j)$.

The non-flow and continuous plug flow systems dealt with in *Chapters 6 and 7* have much in common. A non-flow system can be visualised as a piston-in-cylinder system. Its volume and intensive properties can be made to pass through a time-dependent cycle. In a continuous plug flow system with or without recycle the flow rate and properties pass through a place-dependent cycle. An alternative description as a time-dependent cycle presents itself when following a flow element passing through the system.
The various open and closed cycles are elements of *heat and power engineering*. The discontinuous non-flow and the continuous flow versions of a cycle are described by the same equations. Reversible closed cyles can be operated in a heat engine/power generation mode or a refrigeration mode.

As opposed to non-flow and continuous plug flow systems the properties in continuous mixed flow systems dealt with in *Chapter 8*, are independent of both time and place.

In Sects. 6.2.5 and 8.2.4 we devise for the sake of the exercise systems which accomplish reversible interface transport and reversible reaction. The systems include as *reversibly operating devices* membranes for selective transport of components, Carnot heat engines and pumps to transport heat, open piston-in-cylinder systems for heating/cooling and compression/expansion, and Van 't Hoff boxes for reversible reaction. In each case the net work output of the reversibly operating devices equals the exergy loss for its irreversible counterpart.

Elements of *chemical reaction engineering* consist of the heat balances for a batch reactor, a continuous plug flow reactor and a backmixed reactor in Sects. 6.2.6, 7.2.4 and 8.3.1.

Chapter 9 deals with *physical transport phenomena* and contains elements of *fluid mechanics.* Here properties change both with time and with place in all directions. Rigorous BE's and equations for molecular transport are explored and lead finally to the *set of simplified transport equations* in Sect. 9.3.5. They describe velocity, concentration and temperature as a function of time and place and transfer of momentum, matter and heat at an interface or wall.

1 Principles

All equations in this book are derived from the principles and concepts postulated in this chapter. As indicated in the *structure and contents* the principles are introduced in three consecutive steps. We introduce first the column-1 principles which involve mole numbers, masses and generalised extensives, then the column-2 principles which include the fundamental extensives momentum, energy and entropy, and, finally, the column-3 principles comprising phenomenological laws.

1.1 Column-1 Principles

Throughout the book the systems approach plays a central role. In this cell a system is introduced as a supermolecular amount of matter enclosed within real or fictitious boundaries.

A system has properties. The properties of a system can be divided into intensives and extensives. Intensive properties such as the temperature T and the pressure p specify points of the system, whereas extensive properties such as the control volume V and the component and total mole numbers and masses, n_i, n, m_i and m, specify the whole system and can be expressed as sums of extensives of parts constituting the whole.

Property changes of a system are described by balance equations (BE's). For a generalised extensive E the BE over the system contains terms representing the accumulation of E in the system (the change of E), the net transport of E into the system through its boundaries, and the production of E within the system. The BE [E] takes the form:

$$[E]: \underset{\text{in system}}{\text{accumulation}} = \underset{\text{into system}}{\text{net transport}} + \underset{\text{within system}}{\text{production}}$$

The transport of E through a boundary can be split in transport by matter and immaterial transport, i.e. transport other than by matter. The production terms in the BE's for the mole numbers n_i of components participating in a chemical reaction, are interrelated by the stoichiometric coefficients occurring in the reaction equation. For the total mass m of a system the production is always zero in the absence of nuclear reactions (law of conservation of mass).

When all transport into a system is ceased (in other words when a system is isolated from its environment), property changes of the system will continue to occur as a result of internal production until eventually a state of equilibrium is reached at which all property changes with time and production rates have become zero. Each equilibrium state of the system is completely specified by a limited number of independent properties or degrees of freedom, independently of the path by which that state was reached. Property relations (PR's) connect properties of equilibrium states of a system.

In the special case that a system passes through a series of equilibrium states, stoppage of all transport into the system results instantaneously in zero property changes with time and zero production rates. By reversing the direction of all transport the system returns to a previous equilibrium state. A condition for this to occur is reversibility in sign of all production terms in the describing BE's. The processes proceeding are referred to as reversible. PR's or equilibrium conditions of a system can be found from its describing BE's and reversibility of the production terms.

An equilibrium state of a single-phase system of N components has gradientless intensive properties and 2 + N degrees of freedom, e.g. T, p and N n_i's or m_i's. For a generalised extensive E this leads to the extensive property differential $dE(T, p, n_i)$ with N partial molar properties E_i or to $dE(T, p, m_i)$ with N partial specific properties e_i. The addition rule

$$E = \sum n_i E_i = \sum m_i e_i$$

constitutes a PR between the intensive partial properties associated with E. For the sake of convenience we define as additional single-phase properties the average molar and specific properties $\bar{E}$ and $\bar{e}$, and the concentration and the partial concentration $\check{E}$ and $\check{E}_i$.
For a multi-phase system the average molar and average specific properties $\tilde{\bar{E}}$ and $\tilde{\bar{e}}$ are introduced.

The transport of E by matter through a boundary in a single-phase system can be coupled with the transports of n_i and m_i or the component velocities $\underline{v}_i$ using the local partial properties E_i, e_i or $\check{E}_i$. It can be split in a convective part and a diffusive part due to differences in component velocities $\underline{v}_i$. Thus the diffusion flux of n_i relative to the mole average velocity $\underline{v}^*$, $\underline{\dot{n}}''_{iD}$, and the diffusion flux of m_i relative to the mass average velocity $\underline{v}$, $\underline{\dot{m}}''_{iD}$ present themselves.
The transport of E by matter through a boundary in a multi-phase system can be expressed similarly.

1.2 Column-2 Principles

Here the existence of three fundamental extensives is postulated: momentum $m\underline{v}$, energy $U + X$, the sum of internal and mechanical energy, and entropy S. They are defined by the immaterial transport and production terms in their BE's which are referred to as fundamental BE's (FBE's).

Rates of momentum transport are equivalent with short-range forces. Pressure and shear forces fall in this category.
The rate of production of momentum in a system is equivalent with long-range forces acting on the system such as those by the gravitational field.

Transport of energy is composed of heat δQ and work of short-range forces, δW. The equivalence of heat and work as energy being transported defines heat in terms of work.
The production of energy is constituted by work of long-range forces e.g. work by electrical forces on ions.
All work can be reduced to the scalar product of a force $d\underline{F}$ and a distance $d\underline{z}$ over which that force is exerted.

Finally, we define entropy by its immaterial transport and production terms.
The entropy transport term equals $\delta Q/T$, where T is the local temperature. This transport term leads to an absolute temperature scale when a value is assigned to T at a single specified temperature.
The entropy production δS_p is postulated to be always positive or zero and never negative. It follows that reversibility of the entropy production term is only possible when $\delta S_p = 0$. This condition will be the key to find PR's and equilibrium conditions or BE's for reversible systems from the FBE's.

For the sake of convenience cell 1.2 is concluded by defining a number of auxiliary energy functions: the enthalpy H, the free energy F, the free enthalpy G and the exergy ε in addition to the internal energy U and the mechanical energy X. The partial molars of G are referred to as the chemical potentials μ_i.

1.3 Column-3 Principles

In cell 1.3 the phenomenological laws are introduced as column-3 principles. They involve asymptotic phase behaviour of liquids and gases, and molecular transport of momentum, matter and heat.

The laws on asymptotic phase behaviour are PR's which all liquid and gas mixtures have in common. The infinite-dilution law gives the dependence of the chemical potential μ_i of a component on its mole fraction x_i when x_i approaches zero. Similarly, the ideal-gas law states the dependence of μ_i on the mole fraction y_i and the pressure p when p approaches zero.

The laws on molecular transport are postulated on the basis of the column-2 equation for the volumetric rate of entropy production derived in Sect. 9.2.3,

$$T\dot{S}_p''' = -\underline{\underline{\tau}}: \underline{\nabla}\underline{v} - \sum \dot{\underline{n}}_i'' \cdot \underline{\nabla} A_i - \dot{\underline{Q}}'' . \underline{\nabla}\, T/T,$$

which reflects that each transport flux is down the gradient of its potential: transport of momentum by shear is in the direction of lower velocities, transport of matter is in the direction of a lower diffusion potential, while transport of heat is to a lower temperature.
The transport laws postulated are linear. They are Newton's law for momentum and the Standart–Taylor–Krishna equations for matter and heat.

For a newtonian fluid the momentum-flux tensor depends linearly on the velocity gradient:

$$\underline{\underline{\tau}} = -\mu\{\underline{\nabla}\underline{v} + (\underline{\nabla}\underline{v})^T\}$$

The diffusion flux is a linear combination of the gradients $\underline{\nabla} A_i$ and $\underline{\nabla} T$. In Sect. 9.3.2 we substitute the column-2 equation for $\underline{\nabla} A_i$ as a linear combination of gradients of the chemical potential, the pressure and the electrical potential to obtain the Standart–Taylor–Krishna equation for molecular transport of matter. The rigorous equation describes multicomponent concentration diffusion, pressure diffusion, forced diffusion in an electrical field and thermo-diffusion.

Finally, the equation for molecular transport of heat gives the heat flux as a linear combination of $\underline{\nabla} T$ and the fluxes $\dot{\underline{n}}_i''$. There are two contributions representing heat conduction and the Dufour or diffusion-thermo effect.

1.1.1 Properties and BE's of a System

In this book a system is a supermolecular amount of matter enclosed within real or fictitious boundaries.

The properties of a system are either intensive or extensive. Intensive properties like the temperature T and the pressure p specify a point of a system, whereas extensive properties specify the whole and are equal to the sum of the properties of the constituent parts of a system. The volume V, the component and total mole numbers n_i and n, the component and total masses m_i and m (m_i is coupled with n_i by the molar mass M_i) are extensive properties.

For a generalised extensive E a balance equation (BE) involving accumulation, transport and production terms, is formulated. The transport of E through the boundaries of a system is either by matter or immaterial (i.e. not coupled with transport of matter).

For n_i, n, m_i and m the immaterial transport is by definition always zero. V can change by expansion without transport of matter (closed wall).

The production terms for the mole numbers n_i of components participating in a chemical reaction,

$$0 \rightarrow \sum \nu_i \,\text{mole}\, i,$$

are coupled by the stoichiometric coefficients of the reaction equation. The law of conservation of mass postulates that the production of m is always zero (nuclear reactions are not considered).

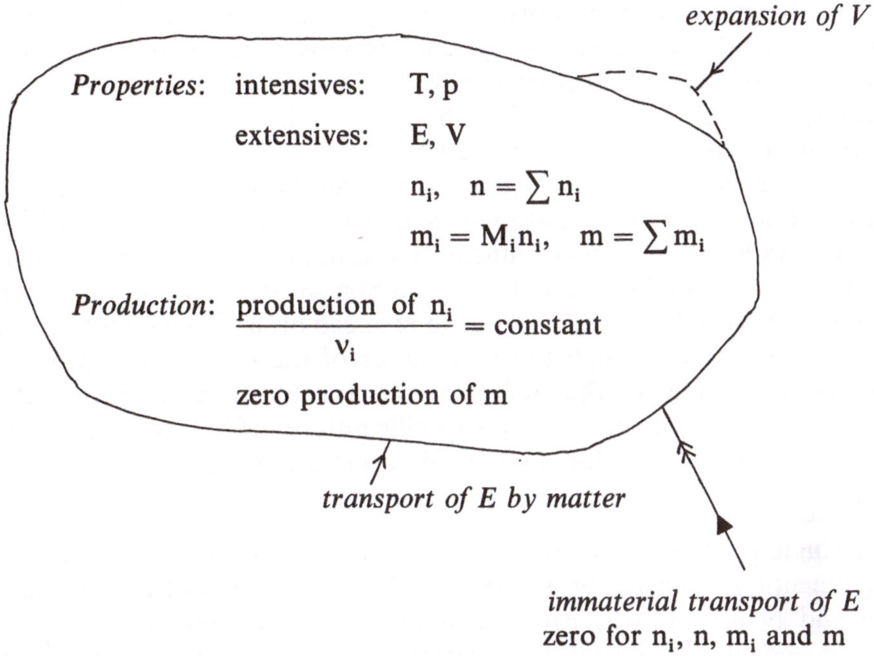

BE of System for E:

[E]:	accumulation of E in system = net transport of E into system + production of E in system

1.1.2 Equilibrium State and PR's of a System

What happens when all transport through the boundaries of a system is ceased, in other words when a system is isolated from its environment? In general the system properties will continue to change as a result of transport and reactions within the system. This will go on until the system reaches eventually an equilibrium state.
In the special case that the system is passing through a series of equilibrium states, the properties stop to change instantaneously upon isolation of the system from its environment. In this limiting case the BE's become PR's interconnecting properties of equilibrium states, while the equilibrium states can be passed through in reversed order by changing the sign of all transport through the boundaries of the system. The processes within the system are then referred to as reversible. An equilibrium state of a system is fully specified by a specific number of independent properties or degrees of freedom and is independent of the path by which that state was reached.

A single-phase system of N components at physical equilibrium has gradientless intensive properties and 2 + N degrees of freedom, e.g. T, p and N n_i's or m_i's. Any extensive E is a function of these 2 + N independents. The coefficients occurring in dE are in turn system properties. The E_i's and e_i's are intensives. Integration of dE at constant intensives yields the addition rule.

An extensive of a multi-phase system is obtained by summation over the constituent phases.

Single-Phase System

E is function of T, p and N n_i's or m_i's

total differential

$$dE = E_T\,dT + E_p\,dp + \sum E_i\,dn_i = E_T\,dT + E_p\,dp + \sum e_i\,dm_i$$

partial molar and partial specific properties

$$E_i = \frac{\partial E}{\partial n_i}, \quad e_i = \frac{\partial E}{\partial m_i}$$

integration of dE at constant intensives

$$E = \sum n_i E_i = \sum m_i e_i$$

addition rule

Multi-Phase System

$$E = \sum_{\alpha} E^{\alpha} = \sum_{\alpha}\sum_{i} n_i^{\alpha} E_i^{\alpha} = \sum_{\alpha}\sum_{i} m_i^{\alpha} e_i^{\alpha}$$

1.1.3 Intensives Associated with Single-Phase Extensives

In addition to the N E_i's and N e_i's, the average molar property $\bar{E}$, the average specific property $\bar{e}$ and the concentration $\check{E}$ are defined as intensives associated with E. These definitions lead in turn to the definition of the mole and mass fractions x_i and w_i, the molar concentrations c_i and c, the mass concentrations or densities ρ_i and ρ and the average molar mass $\bar{M}$. Further the partial concentration $\check{E}_i$, the intensive-state versions of the addition rule and the component volume fraction $\check{V}_i$ present themselves.

Extensive	Associated intensives		
E	$\bar{E} = \frac{E}{n}$	$\bar{e} = \frac{E}{m}$	$\check{E} = \frac{E}{V}$
n_i $n = \sum n_i$ $m_i = M_i n_i$ $m = \sum m_i = \sum M_i n_i$	x_i $\sum x_i = 1$ — $\bar{M} = \sum x_i M_i$	— — w_i $\sum w_i = 1$	c_i $c = \sum c_i$ $\rho_i = M_i c_i$ $\rho = \sum \rho_i$
$n_i = x_i n$ $m_i = w_i m$ $m = \bar{M} n$	— — —	— — —	$c_i = x_i c$ $\rho_i = w_i \rho$ $\rho = \bar{M} c$
$n_i E_i = m_i e_i$ $E = \sum n_i E_i = \sum m_i e_i$ $E = n\bar{E} = m\bar{e}$	— $\bar{E} = \sum x_i E_i$ $\bar{E} = \bar{M}\bar{e}$	— $\bar{e} = \sum w_i e_i$ —	$\check{E}_i = c_i E_i = \rho_i e_i$ $\check{E} = \sum \check{E}_i$ $\check{E} = c\bar{E} = \rho\bar{e}$
$n_i V_i = m_i v_i$ $V = \sum n_i V_i = \sum m_i v_i$ $V = n\bar{V} = m\bar{v}$	— $\bar{V} = \sum x_i V_i$ $\bar{V} = \bar{M}\bar{v}$	— $\bar{v} = \sum w_i v_i$ —	$\check{V}_i = c_i V_i = \rho_i v_i$ $\sum \check{V}_i = 1$ $c\bar{V} = \rho\bar{v} = 1$

1.1.4 Reduced Properties Associated with Multi-Phase Extensives

For a multi-phase system definitions of the reduced multi-phase properties $\tilde{E}$ and $\tilde{e}$ lead to the phase mole and mass fractions x^{α} and w^{α} and the component mole and mass fractions x_i and w_i.

For a two-phase system the phase mole fractions follow from $\bar{E}^{\alpha}$, $\bar{E}^{\beta}$ and $\tilde{E}$ or x_i^{α}, x_i^{β} and x_i by applying the lever rule.

Extensive	Associated reduced properties	
E	$\tilde{E} = \frac{E}{n}$	$\tilde{e} = \frac{E}{m}$
n^α	x^α	—
$n = \sum_\alpha n^\alpha$	$\sum_\alpha x^\alpha = 1$	—
m^α	—	w^α
$m = \sum_\alpha m^\alpha$	—	$\sum_\alpha w^\alpha = 1$
$n_i = \sum_\alpha n_i^\alpha = \sum_\alpha x_i^\alpha n^\alpha$	$x_i = \sum_\alpha x_i^\alpha x^\alpha$	—
$m_i = \sum_\alpha m_i^\alpha = \sum_\alpha w_i^\alpha m^\alpha$	—	$w_i = \sum_\alpha w_i^\alpha w^\alpha$
$E = \sum_\alpha E^\alpha = \sum_\alpha n^\alpha \bar{E}^\alpha$	$\tilde{E} = \sum_\alpha x^\alpha \bar{E}^\alpha$	$\tilde{e} = \sum_\alpha w^\alpha \bar{e}^\alpha$
$= \sum_\alpha m^\alpha \bar{e}^\alpha$		

Level Rule Two-Phase System

$$x^\alpha + x^\beta = 1$$

$$x^\alpha x_i^\alpha + x^\beta x_i^\beta = x_i \qquad x^\alpha (x_i - x_i^\alpha) = x^\beta (x_i^\beta - x_i)$$

$$x^\alpha \bar{E}^\alpha + x^\beta \bar{E}^\beta = \tilde{E} \qquad x^\alpha (\tilde{E} - \bar{E}^\alpha) = x^\beta (\bar{E}^\beta - E)$$

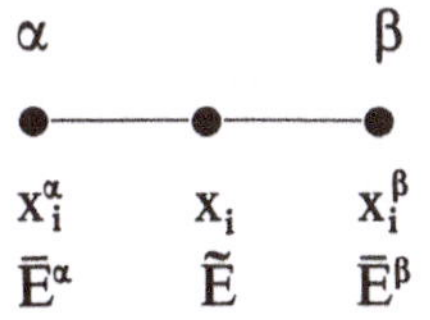

1.1.5 Transport of Extensive E by Matter

The rate of transport of E by matter per unit surface area in a given direction, the transport flux $\underline{\dot{E}}''$, is a vector i.e. a quantity with a magnitude and a direction (see appendix on vectors and tensors). It can be represented by the magnitudes of its projections $\dot{E}_1''$ on three orthogonal coordinate axes with unit vectors $\underline{\delta}_1$ using the convention by which repeated coordinate indices imply summation over the three directions. The transport rate $d\dot{E}$ through an area $d\underline{A}$ is given the scalar product of $\underline{\dot{E}}''$ and $d\underline{A}$ (magnitude dA, direction perpendicular to the surface area). The transport of E through a surface area $d\underline{A}$ in a time interval dt is denoted by $d(\delta E)$.
In the BE's of a system $d\dot{E}$ or $d(\delta E)$ still need to be integrated over the surface of the control volume of the system.

How is the transport of a generalised extensive E by matter coupled with the transport of n_i and m_i? In a single-phase system this is done by assuming local equilibrium and using the addition rule derived in Sect. 1.1.2. In addition to the coupling of $\underline{\dot{E}}''$ with $\underline{\dot{n}}_i''$ and $\underline{\dot{m}}_i''$, $\underline{\dot{E}}''$ can be related to the local component velocities $\underline{v}_i$. In the expressions for $\underline{\dot{n}}''$ and $\underline{\dot{m}}''$ the mole average velocity $\underline{v}^*$ and the mass average velocity $\underline{v}$ appear.

transport flux: $\underline{\dot{E}}'' = \dot{E}_1''\,\underline{\delta}_1 = \dot{E}_1''\,\underline{\delta}_1 + \dot{E}_2''\underline{\delta}_2 + \dot{E}_3''\underline{\delta}_3$

transport rate through area $d\underline{A}$:

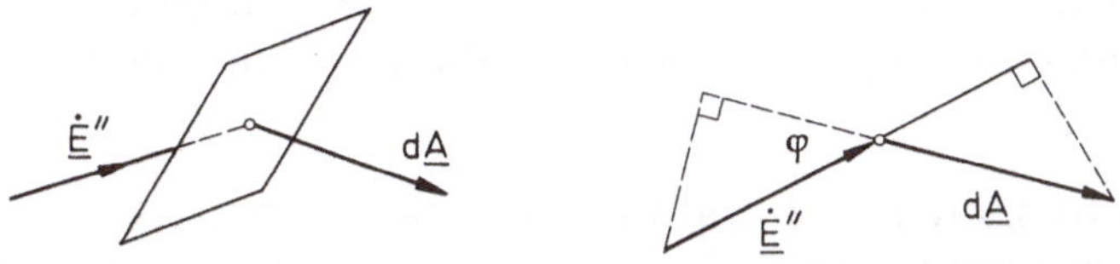

$$d\dot{E} = \underline{\dot{E}}''.\,d\underline{A} = \dot{E}''\,dA\ \cos\varphi = \dot{E}_1''\,dA_1$$

transport of E through $d\underline{A}$ in time interval dt:

$$d(\delta E) = d\dot{E}\,dt = \underline{\dot{E}}''.\,d\underline{A}\ dt$$

Single-phase system

$E = \sum E_i n_i = \sum e_i m_i = \sum \check{E}_i V$

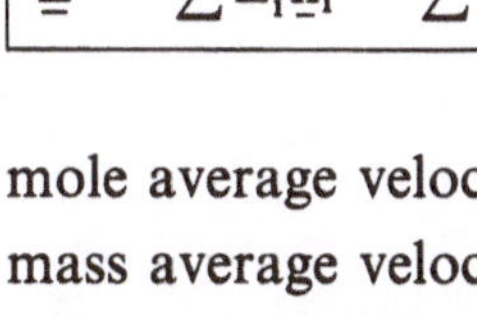

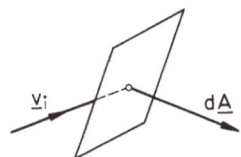

$\underline{\dot{V}}''$ for i transported: $\underline{v}_i$
$d\dot{V}$ for i transported: $\underline{v}_i\,.\,d\underline{A}$
$d(\delta V)$ for i transported: $(\underline{v}_i dt)\,.\,d\underline{A}$

$$\boxed{\underline{\dot{E}}'' = \sum E_i \underline{\dot{n}}_i'' = \sum e_i \underline{\dot{m}}_i'' = \sum \check{E}_i \underline{v}_i}$$

mole average velocity $\underline{v}^* = \sum x_i \underline{v}_i$
mass average velocity $\underline{v} = \sum w_i\, v_i$
volume average velocity $\underline{\dot{V}}'' = \sum \check{V}_i \underline{v}_i$

$$\boxed{\begin{aligned}\underline{\dot{n}}_i'' &= c_i \underline{v}_i\\ \underline{\dot{n}}'' &= \sum \underline{\dot{n}}_i'' = c\underline{v}^*\end{aligned}} \qquad \boxed{\begin{aligned}\underline{\dot{m}}_i'' &= \rho_i \underline{v}_i\\ \underline{\dot{m}}'' &= \sum \underline{\dot{m}}_i'' = \rho\underline{v}\end{aligned}}$$

It is convenient to define the component diffusion fluxes $\underline{\dot{n}}''_{iD}$ relative to $\underline{v}^*$ and $\underline{\dot{m}}''_{iD}$ relative to $\underline{v}$. The definitions lead to a split of $\underline{\dot{n}}''_i$, $\underline{\dot{m}}''_i$ and $\underline{\dot{E}}''$ into convective and diffusive parts. At equal component velocities all diffusion terms vanish. The single-phase mixture is then transported as a whole and local intensive properties can be attributed to the quantities being transported.

The transport of E by matter in a multi-phase system can be coupled with the transport of matter along the same lines. In the absence of diffusion in the individual phases, and at equal phase velocities (no phase slip), the transport equations can be further simplified.

Convection and diffusion in a single-phase system

$$\boxed{\dot{\underline{n}}''_{iD} \equiv c_i(\underline{v}_i - \underline{v}^*)} \qquad \boxed{\dot{\underline{m}}''_{iD} \equiv \rho_i(\underline{v}_i - \underline{v})}$$

$$\sum \dot{\underline{n}}''_{iD} = 0 \qquad \sum \dot{\underline{m}}''_{iD} = 0$$

$$\boxed{\dot{\underline{n}}''_i = c_i\underline{v}^* + \dot{\underline{n}}''_{iD} = x_i\dot{\underline{n}}'' + \dot{\underline{n}}''_{iD}} \qquad \boxed{\dot{\underline{m}}''_i = \rho_i\underline{v} + \dot{\underline{m}}''_{iD} = w_i\dot{\underline{m}}'' + \dot{\underline{m}}''_{iD}}$$

$$\boxed{\begin{aligned}\dot{\underline{E}}'' &= \check{E}\underline{v}^* + \sum E_i\,\dot{\underline{n}}''_{iD} \\ &= \bar{E}\dot{\underline{n}}'' + \sum E_i\,\dot{\underline{n}}''_{iD}\end{aligned}} \qquad \boxed{\begin{aligned}\dot{\underline{E}}'' &= \check{E}\underline{v} + \sum e_i\,\dot{\underline{m}}''_{iD} \\ &= \bar{e}\dot{\underline{m}}'' + \sum e_i\,\dot{\underline{m}}''_{iD}\end{aligned}}$$

$$\dot{\underline{V}}'' = \underline{v}^* + \sum V_i\dot{\underline{n}}''_{iD} \qquad \dot{\underline{V}}'' = \underline{v} + \sum v_i\dot{\underline{m}}''_{iD}$$

Multi-phase system

$$E = \sum\sum E_i^\alpha n_i^\alpha = \sum\sum e_i^\alpha m_i^\alpha$$

transport of E through area $d\underline{A}$ in time dt:

$$d(\delta E) = \sum\sum E_i^\alpha\, d(\delta n_i^\alpha) = \sum\sum e_i^\alpha\, d(\delta m_i^\alpha)$$

no diffusion:

$$d(\delta E) = \sum \bar{E}^\alpha\, d(\delta n^\alpha) = \sum \bar{e}^\alpha\, d(\delta m^\alpha)$$

no diffusion and equal phase velocities:

$$d(\delta E) = \tilde{E}\, d(\delta n) = \tilde{e}\, d(\delta m)$$

1.2.1 Immaterial Transport of Fundamental Extensives

Having completed the introduction of the column-1 principles and definitions, we now turn to the fundamental extensives momentum, energy and entropy.

In Sect. 1.1.1 we formulated a BE of any system for a generalised extensive property E. Three fundamental extensives to which this formalism holds in addition to n_i, n, m_i and m are here postulated to exist: the momentum $m\underline{v}$, the sum of the internal energy and the mechanical energy, U + X, and the entropy S of a system.
The internal energy U and the entropy S are equilibrium-state functions as introduced in Sect. 1.1.2. They occur in PR's and have partial single-phase properties attached to them. In the absence of momentum and mechanical energy effects, BE's equations of a system for U and S can be set up which in the limiting case of reversible processes result in fundamental PR's and equilibrium conditions.
The momentum $m\underline{v}$ and the mechanical energy X do not occur in PR's and have only mass average properties and no partial properties attached to them.

For the fundamental extensives momentum, energy and entropy immaterial transport and production still need to be specified. We start with the immaterial transport denoted by $\dot{\underline{\underline{E}}}''_{im}$, $d\dot{E}_{im}$ and $d(\delta E_{im})$.
Rates of momentum transport are equivalent with short-range forces acting at the boundaries of a system. Pressure and shear forces fall in this category. The momentum flux in the 1-direction is a vector denoted by $\underline{\tau}_l$. It can be written as a product of the m-component of $\underline{\tau}_l$ and the unit vector in the m-direction, $\tau_{lm}\underline{\delta}_m$ in which m is to be summed over the three directions. The momentum-flux or stress tensor $\underline{\underline{\tau}}$ can be expressed in terms of nine components τ_{lm} and nine unit dyads $\underline{\delta}_l\underline{\delta}_m$ (see appendix on vectors and tensors).
Transport of energy is by heat Q or by work W of short-range forces. Work is further discussed in the next section. The equivalence of heat and work as energy being transported defines heat in terms of work. The transport of energy as heat as opposed to the transport of energy as work is usually referred to as the transport of heat. Transport of energy by radiation is beyond the scope of this book.
The immaterial transport of entropy involves heat as well. The transport through an area $d\underline{A}$ in a time interval dt is given by $d(\delta Q)/T$ where T is the locally prevailing temperature. In Sect. 1.1.1 we postulated the existence of T as an intensive property. Later we shall see that the entropy transport term leads to an absolute temperature scale when a value is assigned to T at a single specified temperature. When assigning a value of approx. 273 to melting ice at a pressure of one atmosphere, the Kelvin scale is obtained.

E	$\dot{\underline{E}}''_{im}$	$d\dot{E}_{im} = \dot{\underline{E}}''_{im} . d\underline{A}$	$d(\delta E_{im}) = \dot{\underline{E}}''_{im} . d\underline{A}\ dt$
momentum $m\underline{v}$	fluxes in l-direction 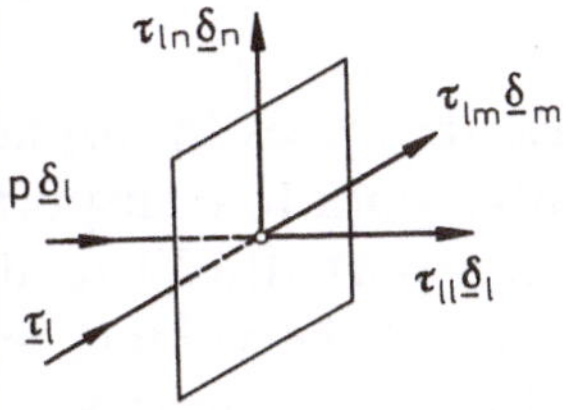	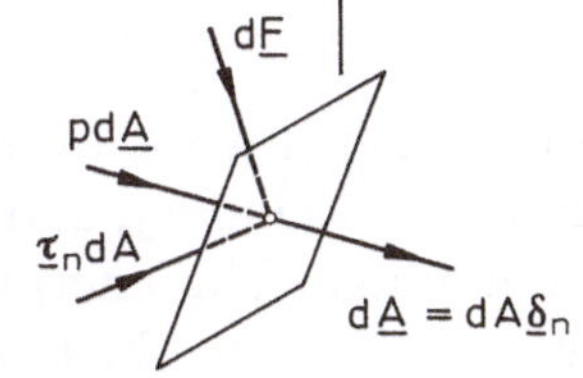	
force pressure shear	 $p\underline{\delta}_l$ $\underline{\tau}_l = \tau_{lm}\,\underline{\delta}_m$	$d\underline{F}$ $p\,d\underline{A}$ $\underline{\tau}_n\ dA$	
	τ_{lm} = m-component of flux in l-direction $\underline{\underline{\tau}} = \tau_{lm}\ \underline{\delta}_l\underline{\delta}_m$ = flux or stress tensor		
energy $U + X$ U = internal energy $X = \frac{1}{2}mv^2 + mgh$ = mechancal energy	$\dot{\underline{Q}}'' + \dot{\underline{W}}''$	$d\dot{Q} + d\dot{W}$	$d(\delta Q) + d(\delta W)$
entropy S	$\dfrac{\dot{\underline{Q}}''}{T}$	$\dfrac{d\dot{Q}}{T}$	$\dfrac{d(\delta Q)}{T}$

1.2.2 Energy Transport by Work

Energy transport through a boundary is composed of heat and work by short-range forces. Here we focus on work.
All work can be related to the scalar product of a force $d\underline{F}$ and a distance $d\underline{z}$ over which that force is exerted.

We shall distinguish work by pressure forces in displacement work to displace a closed boundary, and entry work to transport matter through an open boundary. The energy transport $d(\delta W)$ to displace a closed boundary is opposite to the direction of the volume "exchange" $d(\delta V_x)$ between the two regions separated by the boundary.
The entry work to enter an open boundary is composed of work on the individual components of a single-phase system. For component i the pressure acts on a surface area which is a fraction $\check{V}_i$ of the total surface. The displacements of the components vary with the component velocity $\underline{v}_i$. In the result for $d(\delta W)$ the volume transport $d(\delta V)$ emerges.
The flux of entry work can be decomposed in a convective and a diffusive part (see Sect. 1.1.5). It can alternatively be interpreted as the flux of displacement work for a boundary moving with the mass average $\underline{v}$ and a superimposed flux of entry work for the components moving with the relative velocity $\underline{v}_i - \underline{v}$ through the boundary.

Work by shear forces is another category of energy transport. The flux of work $\underline{\dot{W}}''$ by shear forces is the single-dot product of $\underline{\underline{\tau}}$ and $\underline{v}$. The rate of work $d\dot{W}$ by shear forces on an area $d\underline{A}$ involves the scalar product of the momentum flux $\underline{\tau}_n$ and the velocity $\underline{v}$.

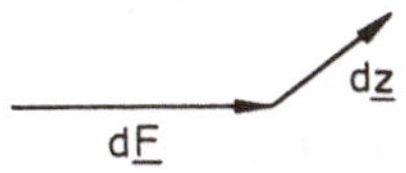 *work by a force*	$\boxed{d(\delta W) = d\underline{F}.\, d\underline{z}}$ $d\dot{W} = d\underline{F}.\underline{v}$	
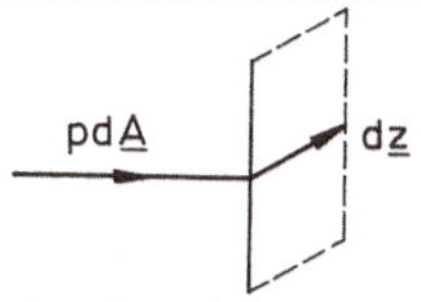 *displacement work*	$d(\delta W) = p d\underline{A}.\, d\underline{z}$ $\boxed{d(\delta W) = -\, p d(\delta V_x)}$	$d(\delta V_x) = -\, d\underline{A}.\, d\underline{z}$
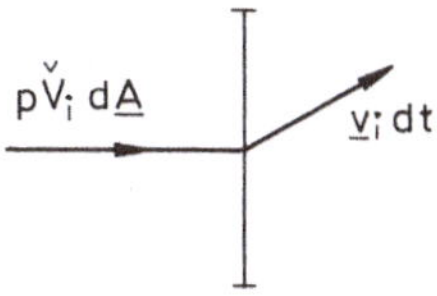 *entry work*	$d(\delta W) = \sum p\, \check{V}_i\, d\underline{A}.\, \underline{v}_i\, dt$ $\boxed{d(\delta W) = p\, d(\delta V)}$ $d\dot{W} = p\, d\dot{V}$ $\underline{\dot{W}}'' = p\, \underline{\dot{V}}''$ $\boxed{\underline{\dot{W}}'' = p\, \underline{v} + \sum p\, v_i\, \underline{\dot{m}}''_{iD}}$	$d(\delta V) = \sum \check{V}_i\, \underline{v}_i . d\underline{A}\, dt$
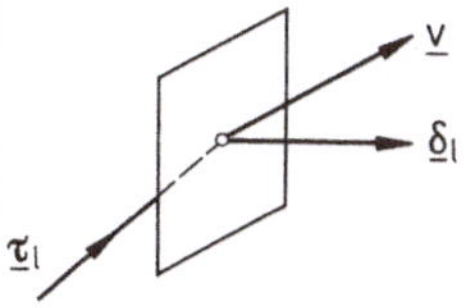 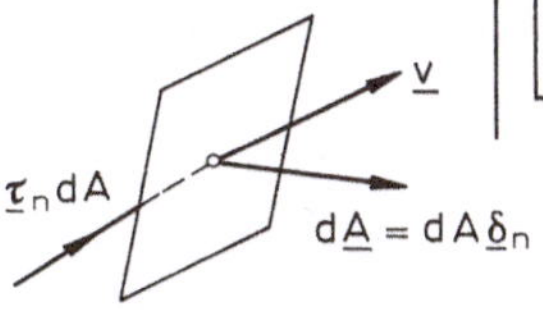 *work by shear forces*	$\underline{\dot{W}}'' = \underline{\tau}_l.\, \underline{v}\, \underline{\delta}_l = \tau_{lm}\, v_m\, \underline{\delta}_l$ $\boxed{\underline{\dot{W}}'' = \underline{\underline{\tau}}.\, \underline{v}}$ $\boxed{d\dot{W} = \underline{\tau}_n.\, \underline{v}\, dA}$	

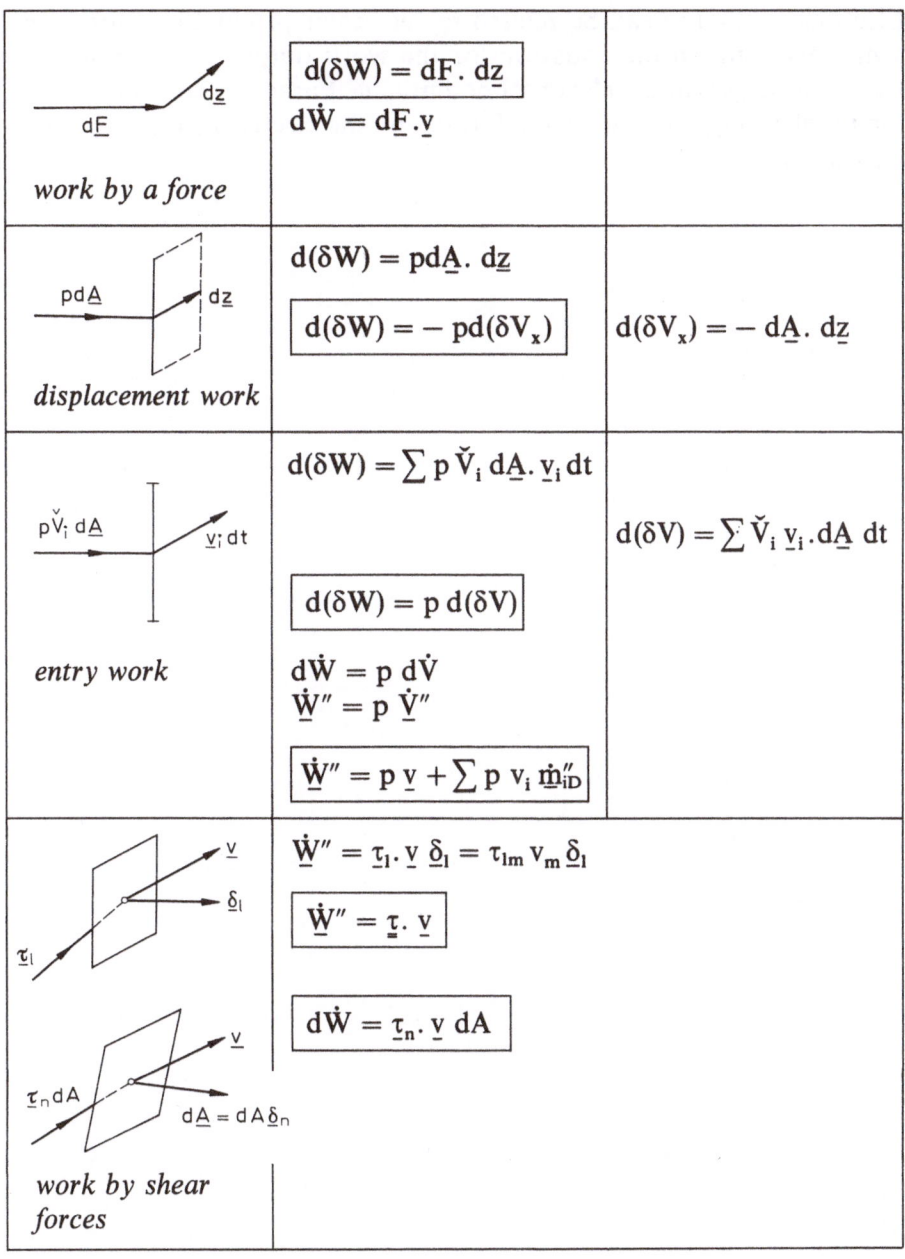

Electrical work too can be related to the scalar product of a force and a displacement. In the equation for the work output δW^- of an electrochemical system the electrical potential ϕ, Faraday's constant $\mathscr{F}$ and the number δn_{kp} of equivalents formed by the electrochemical reaction, play a role.

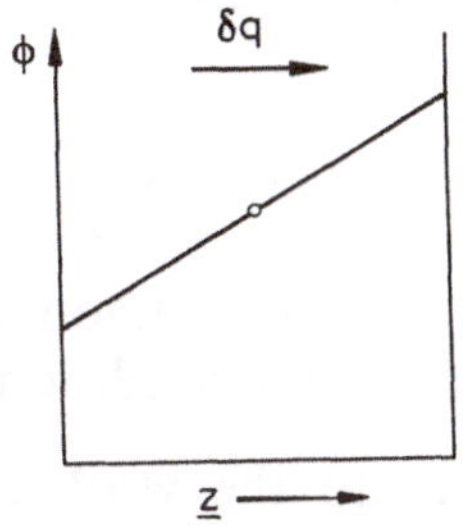

$$d(\delta W) = (\underline{\nabla}\phi\, \delta q) . d\underline{z}$$

$$\boxed{\delta W = \Delta\phi\, \delta q}$$

electrical work

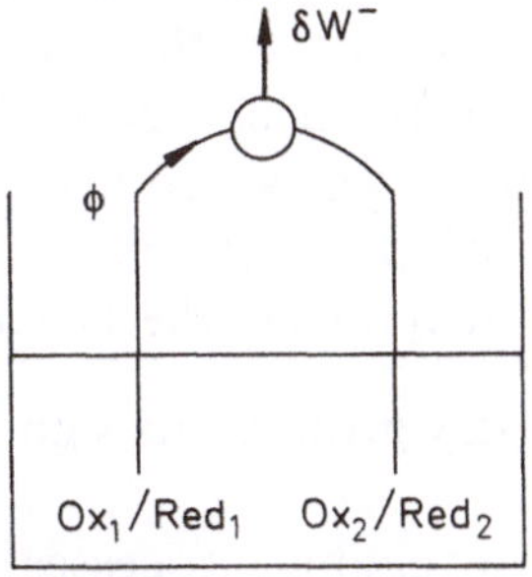

$$\boxed{\delta W^{-} = \phi\, \mathscr{F}\, \delta n_{kp}}$$

$$\mathscr{F} = 0.965 * 10^{8}\ \mathrm{C.\ (kg\ equivalent)^{-1}}$$

$$\nu_{ox_1}\, Ox_1 \rightarrow \nu_{Red_1}\, Red_1 + 1\oplus$$

$$\nu_{Red_2}\, Red_2 + 1\oplus \rightarrow \nu_{Ox_2}\, Ox_2$$

$$\nu_{Ox_1}\, Ox_1 + \nu_{Red_2}\, Red_2 \rightarrow \nu_{Red_1}\, Red_1 + \nu_{Ox_2}\, Ox_2$$

electrochemical work

1.2.3 Production of Fundamental Extensives. Energy Functions

We now focus on the rates of production per unit volume $\dot{E}_p'''$ of momentum, energy and entropy.

The production of momentum is equivalent to long-range forces. We consider only the gravitational field denoted by $\underline{g}$, the force per unit mass. The net force exerted by an electrical field on a system with ions is zero when the net electrical charge is zero (electroneutrality condition).

The production of energy in a system is equivalent to the work by long-range forces. We choose to incorporate the potential energy in the gravitational field, mgh, as part of the total energy of a system. This choice makes the production of energy by the gravitational field zero. The work by an electrical field on ions does contribute to the energy production in a system. The force per unit of electrical charge is the negative of the gradient of the electrical potential, $-\underline{\nabla}\phi$ where $\underline{\nabla}$ is the vector differential operator (nabla). The rate of work per unit volume by this force is readily found as a sum over the individual ions in the system.

Finally, the production of entropy in a system is only given by its sign. It is positive or zero, but never negative.
In Sect. 1.1.2 we discussed the limiting case of reversible processes in a system which passes through a series of equilibrium states. To pass through the equilibrium states in reversed order, all terms in the fundamental BE's must have reversible signs. With an entropy production term which is always positive or zero, this is only possible when the entropy production is zero.

For reasons of convenience a number of energy functions is defined in addition to the internal energy U: the enthalpy H, the free energy F, the free enthalpy G and the exergy ε. The partial molar free enthalpies are referred to as the chemical potentials μ_i.

E	$\dot{E}_p'''$ production rate per unit volume
momentum $m\underline{v}$ by gravitational field $\underline{g}$	$\boxed{\rho \underline{g}}$
energy $U + X$ by electrical field $-\frac{\partial \phi}{\partial x_1}\underline{\delta}_1 = -\underline{\nabla}\phi$ $\underline{\check{F}}_i = -\underline{\nabla}\phi\, c_i z_i \mathscr{F}$ $\sum c_i z_i = 0\,,\quad \sum \underline{\check{F}}_i = 0$	$\sum \underline{\check{F}}_i \cdot \underline{v}_i = \sum \check{F}_i \cdot (\underline{v}_i - \underline{v})$ $= \sum w_i \rho \underline{f}_i \cdot (\underline{v}_i - \underline{v})$ $\boxed{\sum \underline{\check{F}}_i \cdot \underline{v}_i = \sum \underline{\dot{m}}_{iD}'' \cdot \underline{f}_i}$
entropy S	$\boxed{\dot{S}_p''' \geqslant 0}$

Energy functions

enthalpy	$H = U + pV$	
free energy	$F = U - TS$	
free enthalpy	$G = U + pV - TS$	chemical potential $\mu_i \equiv G_i$
exergy	$\varepsilon = H - T^*S$	

1.3.1 Asymptotic Phase Behaviour

We have now arrived at the column-3 principles, the phenomenological laws on asymptotic phase behaviour and molecular transport of momentum, heat and matter.

The laws on asymptotic phase behaviour are PR's which all liquid and gas mixtures have in common. They represent experimental laws reduced under the light of the preceding principles. The infinite-dilution law gives the dependence of the chemical potential μ_i on its mole fraction x_i when x_i approaches zero, while the ideal-gas law gives the dependence of μ_i on the mole fraction y_i and the pressure p when p approaches zero. In both laws the universal gas constant R occurs.

infinite-dilution law:

$$(d\mu_i)_{T,p} = RT\, d(\ln x_i) \quad \text{if } x_i \to 0$$

ideal-gas law:

$$(d\mu_i)_T = RT\, d\ln(y_i p) \quad \text{if } p \to 0$$

R = universal gas constant, 8.3143 $kJ.kmole^{-1}.K^{-1}$

1.3.2 Transport of Momentum

In 9.2.3 an expression is derived for the entropy production $\dot{S}_p'''$ due to the transport fluxes of momentum, matter and heat, $\underline{\underline{\tau}}$, $\underline{\dot{n}}_i''$ and $\underline{\dot{Q}}''$. In the expression for $T\dot{S}_p'''$ each flux is accompanied by a gradient, $\underline{\underline{\tau}}$, by the velocity gradient $\underline{\nabla}\underline{v}$, $\underline{\dot{n}}_i''$ by the gradient of the diffusion potential $\underline{\nabla}A_i$ and $\underline{\dot{Q}}''$ by the temperature gradient $\underline{\nabla}T$. The expression is derived from column-1 and column-2 principles and definitions, and one additional postulate by which the entropy production due to diffusion is invariant to the choice of the reference velocity $\underline{v}_{ref}$. In 9.2.3 an expression for $\underline{\nabla}A_i$ in terms of known quantities is derived as well.

Here we postulate linear experimental transport laws which give the fluxes as linear functions of the gradients as appearing in the expression for $T\dot{S}_p'''$.

We start with the linear momentum transport law which holds for newtonian fluids. The momentum-flux tensor $\underline{\underline{\tau}}$ has as a proportionality constant the dynamic viscosity μ. The expression contains in addition to $\underline{\nabla}\underline{v}$ the transpose of $\underline{\nabla}\underline{v}$ with interchanged coordinate indices. Further it contains the unit tensor $\underline{\underline{\delta}}$ of which the elements are given by the Kronecker delta δ_{lm} to be 1 at equal and 0 at unequal coordinate indices. The momentum-flux tensor $\underline{\underline{\tau}}$ as given by the linear transport law is symmetric: the values of its elements do not change when the coordinate indices are interchanged.
Combination of the expression for $T\dot{S}_p'''$ and the linear transport law reveals that the dynamic viscosity μ is a positive property.

Entropy Production $T\dot{S}_p''' = -\underline{\underline{\tau}}:\nabla\underline{v} - \sum \underline{\dot{n}}_i''.\nabla A_i - \dot{Q}''.\frac{\nabla T}{T}$

$$\underline{\dot{n}}_i'' = x_i c(\underline{v}_i - \underline{v}_{ref})$$

$$\sum x_i \nabla A_i = 0 \quad \text{(invariance to } \underline{v}_{ref})$$

Entropy production due to momentum flux

$$T\dot{S}_p''' = -\underline{\underline{\tau}}:\nabla\underline{v}$$

Linear momentum transport law

$$\underline{\underline{\tau}} = -\mu\{\nabla\underline{v} + (\nabla\underline{v})^T - \tfrac{2}{3}(\nabla.\underline{v})\,\underline{\underline{\delta}}\}$$

$$\tau_{lm} = -\mu\left\{\frac{\partial}{\partial x_l}v_m + \frac{\partial}{\partial x_m}v_l - \tfrac{2}{3}(\nabla.\underline{v})\,\delta_{lm}\right\}$$

Combination of $T\dot{S}_p'''$ and linear momentum transport law

$$\begin{aligned}
\underline{\underline{\tau}}:\underline{\underline{\tau}} &= \underline{\underline{\tau}}:[-\mu\{\nabla\underline{v} + (\nabla\underline{v})^T - \tfrac{2}{3}(\nabla.\underline{v})\,\underline{\underline{\delta}}\}]\\
&= -2\mu(\underline{\underline{\tau}}:\nabla\underline{v}) - \tfrac{2}{3}\mu(\nabla.\underline{v})\,\underline{\underline{\tau}}:\underline{\underline{\delta}}\\
&= -2\mu(\underline{\underline{\tau}}:\nabla\underline{v}) - \tfrac{2}{3}\mu(\nabla.\underline{v})\left[-\mu\left\{2\frac{\partial}{\partial x_l}v_l - \tfrac{2}{3}(\nabla.\underline{v})\,\delta_{ll}\right\}\right]\\
&= -2\mu(\underline{\underline{\tau}}:\nabla\underline{v}) - \tfrac{2}{3}\mu(\nabla.\underline{v})[-\mu\{2(\nabla.\underline{v}) - \tfrac{2}{3}(\nabla.\underline{v})3\}]\\
&= -2\mu(\underline{\underline{\tau}}:\nabla\underline{v})
\end{aligned}$$

$$T\dot{S}_p''' = -\underline{\underline{\tau}}:\nabla\underline{v} = \frac{\underline{\underline{\tau}}:\underline{\underline{\tau}}}{2\mu} = \frac{\tau_{lm}\tau_{ml}}{2\mu} = \frac{\sum\sum\tau_{lm}^2}{2\mu} > 0$$

$\underline{\mu > 0}$

1.3.3 Transport of Matter and Heat

The entropy production due to the transport of matter can be decomposed in products of fluxes and accompanying gradients for diffusion of components relative to each other. The gradients for the diffusion of i relative to j and vice versa have opposite signs: $\underline{\nabla} A_{ij} = -\underline{\nabla} A_{ji}$.

The linear transport laws involve fluxes and gradients as occurring in $T\dot{S}_p'''$ for the diffusion of the component pairs and for the transport of heat. In general the flux for each transport process is a linear combination of its accompanying gradient and the gradients for the other transport processes. Any two transport processes influence each others fluxes via equal coefficients of their gradients (Onsager's reciprocity principle). The mutual influence of the diffusion of one pair of components on that of another pair is postulated to be zero (zero cross-coefficients).

Entropy production $T\dot{S}_p''' = -\sum \underline{\dot{n}}_i'' . \underline{\nabla} A_i - \underline{\dot{Q}}'' . \frac{\underline{\nabla} T}{T}$

$$\underline{\dot{n}}_i'' = x_i c(\underline{v}_i - \underline{v}_{ref}) , \quad \sum x_i \underline{\nabla} A_i = 0$$

Reformulation of entropy production

$$E_i \equiv \sum x_j E_{ij}$$

$$\sum x_i E_i = \tfrac{1}{2}(\sum x_i E_i + \sum x_j E_j) = \tfrac{1}{2}\sum_i \sum_j x_i x_j (E_{ij} + E_{ji})$$

$\sum x_i E_i = 0$ implies $E_{ji} = -E_{ij}$ and vice versa.

$$\sum \underline{\dot{n}}_i'' . \underline{\nabla} A_i = \tfrac{1}{2}(\sum \underline{\dot{n}}_i'' . \underline{\nabla} A_i + \sum \underline{\dot{n}}_j'' . \underline{\nabla} A_j) = \tfrac{1}{2}\sum_i \sum_j (x_j \underline{\dot{n}}_i'' - x_i \underline{\dot{n}}_j'')\underline{\nabla} A_{ij}$$

$$\boxed{T\dot{S}_p''' = -\sum_i \sum_j (\underline{v}_i - \underline{v}_j) . \{\tfrac{1}{2} x_i c x_j \underline{\nabla} A_{ij}\} - \underline{\dot{Q}}'' . \frac{\underline{\nabla} T}{T}}$$

new fluxes: $\underline{v}_i - \underline{v}_j$

new gradients: $\tfrac{1}{2} x_i c x_j \underline{\nabla} A_{ij}$ $\quad \sum_j x_j \underline{\nabla} A_{ij} = \underline{\nabla} A_i$

$$\sum_i x_i \underline{\nabla} A_i = 0 , \quad \underline{\nabla} A_{ij} = -\underline{\nabla} A_{ji}$$

Linear transport laws

$$\boxed{\underline{v}_i - \underline{v}_j = -D_{ij} \frac{\underline{\nabla} A_{ij}}{RT} - D_{ij}\,\alpha_{ij} \frac{\underline{\nabla} T}{T}}$$

$$\boxed{\underline{\dot{Q}}'' = -\lambda \underline{\nabla} T - \sum \sum D_{ij}\,\alpha_{ij}(\tfrac{1}{2} x_i c x_j \underline{\nabla} A_{ij})}$$

In the linear law for transport of matter the Maxwell–Stefan diffusivity D_{ij} and the thermal diffusion factor α_{ij} occur. For a pair of components $D_{ji} = D_{ij}$ and $\alpha_{ji} = -\alpha_{ij}$.
Rearrangement and summation over j yield the final form of the linear law for transport of matter. Together with the expression for $\underline{\nabla} A_i$ derived in 9.2.3. it describes multi-component diffusion driven by concentration and pressure gradients and by electrical forces as well as thermo-diffusion.

The linear law for transport of heat can be rearranged to a form in which the conductivity λ appears as a coefficient. Further the law describes the Dufour or diffusion-thermo effect.

Linear law for transport of matter

$$\underline{v}_i - \underline{v}_j = -D_{ij}\frac{\underline{\nabla}A_{ij}}{RT} - D_{ij}\,\alpha_{ij}\frac{\underline{\nabla}T}{T}$$

$$\underline{v}_j - v_i = -(\underline{v}_i - \underline{v}_j)$$

$$\underline{\nabla}A_{ji} = -\underline{\nabla}A_{ij}$$

$$D_{ji} = D_{ij} \quad \text{(Maxwell–Stefan diffusivities)}$$

$$\alpha_{ji} = -\alpha_{ij} \quad \text{(thermal diffusion factors)}$$

$$\frac{x_j\dot{\underline{n}}_i'' - x_i\dot{\underline{n}}_j''}{cD_{ij}} = -x_i x_j\frac{\underline{\nabla}A_{ij}}{RT} - x_i x_j \alpha_{ij}\frac{\underline{\nabla}T}{T}$$

$$\boxed{\sum_j \frac{x_j\dot{\underline{n}}_i'' - x_i\dot{\underline{n}}_j''}{cD_{ij}} = -x_i\frac{\underline{\nabla}A_i}{RT} - x_i\alpha_i\frac{\underline{\nabla}T}{T}}$$

$$\alpha_i = \sum_j x_j \alpha_{ij}$$

$$\sum_i x_i \alpha_i = 0$$

Linear law for transport of heat

$$\underline{\nabla}A_{ij} = -\frac{RT}{D_{ij}}(\underline{v}_i - \underline{v}_j) - RT\,\alpha_{ij}\frac{\underline{\nabla}T}{T}$$

$$\dot{\underline{Q}}'' = -\lambda'\underline{\nabla}T - \sum\sum D_{ij}\,\alpha_{ij}(\tfrac{1}{2}x_i c x_j\,\underline{\nabla}A_{ij})$$

$$= -\lambda\underline{\nabla}T + \tfrac{1}{2}RT\sum\sum(x_j\dot{\underline{n}}_i'' - x_i\dot{\underline{n}}_j'')\,\alpha_{ij}$$

$$\boxed{\dot{\underline{Q}}'' = -\lambda\underline{\nabla}T + RT\sum \alpha_i\dot{\underline{n}}_i''}$$

Finally, combination of $T\dot{S}_p'''$ and the linear transport laws reveals that D_{ij} and λ are positive properties.

Combination of $T\dot{S}_p'''$ and linear transport laws

$$T\dot{S}_p''' = -\sum\sum(\underline{v}_i - \underline{v}_j).\{\tfrac{1}{2}x_i c x_j \underline{\nabla} A_{ij}\} - \underline{\dot{Q}}''.\frac{\underline{\nabla}T}{T}$$

$$= -\sum\sum(\underline{v}_i - \underline{v}_j).\left[\tfrac{1}{2}x_i c x_j\left\{-\frac{RT}{D_{ij}}(\underline{v}_i - \underline{v}_j) - RT\alpha_{ij}\frac{\underline{\nabla}T}{T}\right\}\right]$$

$$-(-\lambda\underline{\nabla}T + RT\sum\alpha_i\underline{\dot{n}}_i'').\frac{\underline{\nabla}T}{T}$$

$$\boxed{T\dot{S}_p''' = \tfrac{1}{2}cRT\sum\sum\frac{x_i x_j(\underline{v}_i - \underline{v}_j)^2}{D_{ij}} + \frac{\lambda(\underline{\nabla}T)^2}{T}}$$

$$\underline{D_{ij} > 0},\ \underline{\lambda > 0}$$

2 First Deductions

In this chapter we deduce some first results from the column-1 and column-2 principles postulated in Chapter 1 to be applied in deriving PR's and BE's in the next chapters.

2.1 Column-1 First Deductions

The BE's postulated in cell 1.1 are here simply reformulated to BE's over a differential time interval dt and to BE's per unit time. The BE's include [E] for a generalised extensive, [V], the mole balances [n_i] and [n], and the mass balances [m_i] and [m].

2.2 Column-2 First Deductions

Here we incorporate the immaterial transport through a boundary and the production in a system of the fundamental extensives as postulated in cell 1.2, to formulate the FBE's [U] and [S] over a time interval and the FBE's [$m\underline{v}$], [U + X] and [S] per unit time.

To formulate the exergy balance we apply the FBE's to a system which through part of its boundaries can exchange components and heat with an infinite reservoir which has temperature T*, pressure p* and a set of components in abundant supply.
By combining the FBE's [U] and [S] or [U + X] and [S] over this system, the exergy $\varepsilon = H - T^*S$ presents itself as an extensive property. In the exergy balance of a system exergy transport by heat and exergy transport by matter occur as equivalents of work. Heat available at the temperature level T* has zero exergy, while components in abundant supply have zero partial exergies under the conditions prevailing in the infinite reservoir.
Loss of exergy or potential work can be related to the entropy production due to irreversible processes proceeding in the system and written as $T^* \delta S_p$ or $T^*\dot{S}_p$.

Equilibrium conditions are derived by applying [U] and [S] to suitably devised systems and equating the calculated entropy production $\delta S_p = 0$.

The expression for δS_p due to interfacial transport of energy, interface displacement and transport of matter across an interface reveals that energy is always transported in the direction of lower T, an interface is displaced in the direction of lower p and transport of n_i is to regions of lower μ_i.

The interface equilibrium conditions follow from $\delta S_p = 0$: T, p and the μ_i's are gradientless when crossing the interface.

For a single-phase system or a multi-phase system having gradientless T, p and μ_i's, with reversible boundary displacement and reversible transport of matter and heat through its boundaries, we find the fundamental property relation (FPR) of matter,

$$dU = TdS - pdV + \sum \mu_i dn_i .$$

The FPR of matter can be interpreted as a condition for intraphase physical equilibrium.

For a system at physical equilibrium the entropy production due to a chemical reaction is

$$\delta S_p = -\sum \nu_i \mu_i \, \delta n_{kp}/T = -\Delta \bar{G} \, \delta n_{kp}/T$$

The chemical reaction can be seen to proceed in the direction of lower free enthalpy G. From $\delta S_p = 0$ we find the condition for chemical equilibrium,

$$\sum \nu_i \mu_i = \Delta \bar{G} = 0$$

Similarly the condition for electrochemical equilibrium is found to be

$$e = -\sum \nu_i \mu_i / \mathcal{F} = -\Delta \bar{G} / \mathcal{F}$$

2.1.1 BE's Over a Time Interval and per Unit Time

The BE of a system over a time interval dt for a generalised extensive E comprises the accumulation term dE, the net transport into the system $-\Delta d(\delta E_x)$ and the production of E, δE_p. The transport of E through an incremental part of its boundary, $d(\delta E_x)$ is composed of transport by matter, $d(\delta E)$ and of immaterial transport, $d(\delta E_{im})$. Integration of the incremental inputs over the surface area of the control volume of the system is denoted by $-\Delta$.
The balances for n_i, n, m_i and m are interdependent. The production of n_i by a reaction is here coupled with the production of a component k, δn_{kp}, for which the stoichiometric coefficient is chosen to be $+1$. The law of conservation of total mass m (zero production of m) results in a relation between the molar masses M_i of the components participating in the reaction.

Analogous BE's per unit time rather than over a time interval involve transport and production rates.

BE's over a time interval

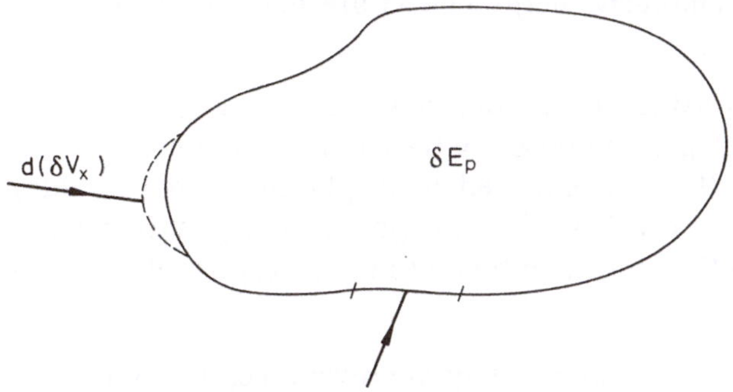

E	$dE = -\Delta d(\delta E_x) + \delta E_p$	$d(\delta E_x) = d(\delta E) + d(\delta E_{im})$	
V	$dV = -\Delta d(\delta V_x)$		
n_i	$dn_i = -\Delta d(\delta n_i) + \nu_i \delta n_{kp}$		$\nu_k \equiv +1$
$n = \sum n_i$	$dn = -\Delta d(\delta n) + \sum \nu_i \delta n_{kp}$	$d(\delta n) = \sum d(\delta n_i)$	
$m_i = M_i n_i$	$dm_i = -\Delta d(\delta m_i) + M_i \nu_i \delta n_{kp}$	$d(\delta m_i) = M_i d(\delta n_i)$	
$m = \sum m_i$	$dm = -\Delta d(\delta m)$	$d(\delta m) = \sum d(\delta m_i)$	$\sum \nu_i M_i = 0$

BE's per unit time

[E]: $\frac{dE}{dt} = -\Delta(d\dot{E}_x) + \dot{E}_p$ $\qquad$ $d\dot{E}_x = d\dot{E} + d\dot{E}_{im}$

2.2.1 FBE's Over a Time Interval

We are now ready to formulate the FBE's of a system over a time interval dt. Momentum and mechanical energy effects are here assumed to be zero.

The production of U is zero in the absence of work by long-range forces. The transport of U, $d(\delta U_x)$, is composed of transport by matter, heat and work. Work can be further decomposed in displacement work, entry work and any other work $d(\delta W')$. The transport of U by matter $d(\delta U)$ and the entry work $d(p\delta V)$ can be combined to the transport of enthalpy by matter $d(\delta H)$.

When the processes proceeding in a system are reversible, the signs of all terms occurring in the FBE's must be reversible. As δS_p is always positive or zero, this implies that δS_p is zero for the limiting case of reversible processes.

FBE's over a time interval

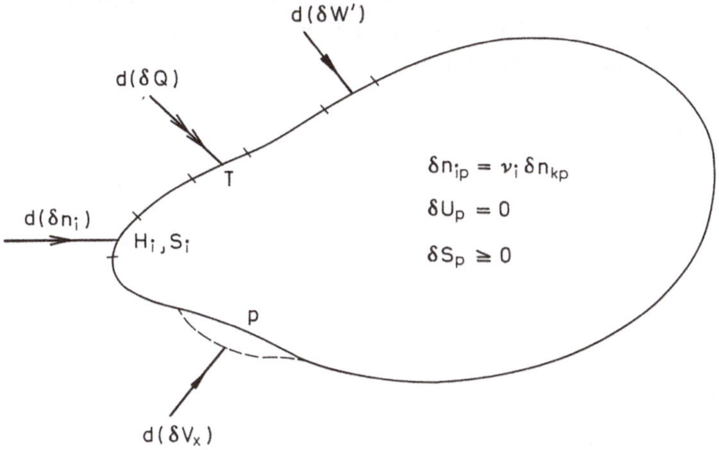

[E]	$dE = -\Delta d(\delta E_x) + \delta E_p$	$d(\delta E_x) = d(\delta E) + d(\delta E_{im})$ $d(\delta E) = \sum E_i\, d(\delta n_i)$
[V] $[n_i]$ [U]	$dV = -\Delta d(\delta V_x)$ $dn_i = -\Delta\, d(\delta n_i) + \nu_i\, \delta n_{kp}$ $dU = -\Delta d(\delta U_x)$	$d(\delta U_x) = d(\delta U) + d(\delta Q) + d(\delta W_t)$ $= d(\delta U) + d(\delta Q) - pd(\delta V_x) + pd(\delta V) + d(\delta W')$ $d(\delta U_x) = d(\delta H) + d(\delta Q) - pd(\delta V_x) + d(\delta W')$
[S]	$dS = -\Delta d(\delta S_x) + \delta S_p$	$d(\delta S_x) = d(\delta S) + \dfrac{d(\delta Q)}{T}$

2.2.2 FBE's per Unit Time

At this stage the BE's of a system per unit time for m, m_i, $m\underline{v}$, U + X and S can readily be written down. For $m\underline{v}$ and X the transport coupled with matter is only by convection. In [U + X] $d\dot{W}$ is the rate of work other than entry work. The latter is incorporated in $d\dot{H}$.

[E]	$\boxed{\dfrac{dE}{dt} = -\Delta(d\dot{E} + d\dot{E}_{im}) + \dot{E}_p}$ $d\dot{E} = \sum E_i\, d\dot{n}_i = \sum e_i\, d\dot{m}_i$ $= \bar{E}\, d\dot{n} + \sum E_i\, d\dot{n}_{iD} = \bar{e} d\dot{m} + \sum e_i\, d\dot{m}_{iD}$
[m]	$\boxed{\dfrac{dm}{dt} = -\Delta(d\dot{m})}$
$[m_i]$	$\boxed{\dfrac{dm_i}{dt} = -\Delta(w_i\, d\dot{m} + d\dot{m}_{iD}) + M_i \nu_i \dot{n}_{kp}}$
$[m\underline{v}]$	$\boxed{\dfrac{d}{dt}(m\underline{v}) = -\Delta(\underline{v}\, d\dot{m} + d\underline{F} + p d\underline{A} + \underline{\tau}_n\, dA) + m\underline{g}}$
[U + X]	$\dfrac{d}{dt}(U + X) = -\Delta(d\dot{U} + d\dot{X} + d\dot{Q} + p d\dot{V} + d\dot{W})$
$\bar{x} = \frac{1}{2}v^2 + gh$	$\boxed{\dfrac{d}{dt}(U + X) = -\Delta(d\dot{H} + \bar{x}\, d\dot{m} + d\dot{Q} + d\dot{W})}$
[S]	$\boxed{\dfrac{dS}{dt} = -\Delta\left(d\dot{S} + \dfrac{d\dot{Q}}{T}\right) + \dot{S}_p}$

2.2.3 Exergy

We consider a system with possible exchange of matter and heat through part of its boundaries with an infinite reservoir. In practical terms such a reservoir could be constituted by surrounding air, water and solid compounds in abundant supply. The infinite reservoir is specified by T^*, p^* and the composition of a set of compounds j in abundant supply.

In the BE's of this system over a time interval dt for U and S the heat transport $d(\delta Q^*)$ occurs. The compounds j do not contribute to $d(\delta H)$ and $d(\delta S)$ when we choose H_j^* and S_j^* to be zero.

The two FBE's can be combined to give the exergy or potential work balance of a system. In this balance exergy transport by matter and by heat occur as equivalents of work. $T^*\delta S_p$ represents the exergy loss due to irreversible processes in the system. It equals the extra heat absorbed from the infinite reservoir and converted into work output if all processes were reversible. For reversible processes the exergy loss is zero and, consequently, the heat absorbed from the infinite reservoir and the work output are maximal (heat rejected and work input are minimal).

The exergy balance is particularly useful for a continuous system with time-independent properties and rates. For such a system the accumulation terms in the BE's vanish and the sum of net inputs of exergy by matter, heat and work equals the exergy loss $T^*\delta S_p$.

The exergy balance per unit time is readily found along the same lines. This time mechanical energy is included. The rate of exergy loss amounts to $T^*\dot{S}_p$.

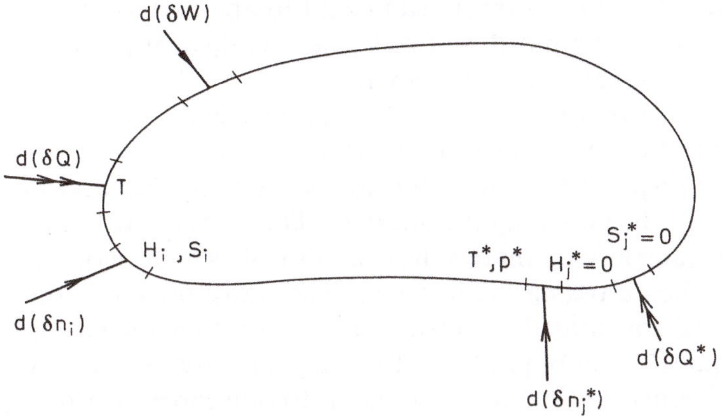

BE's over a time interval

[U]: $$dU = -\Delta\{d(\delta H) + d(\delta Q) + d(\delta Q^*) + d(\delta W)\}$$

[S]: $$dS = -\Delta\left\{d(\delta S) + \frac{d(\delta Q)}{T} + \frac{d(\delta Q^*)}{T^*}\right\} + \delta S_p$$

[U] − T*[S]:

$$\boxed{d(U - T^*S) = -\Delta\left\{d(\delta\varepsilon) + d(\delta Q)\left(1 - \frac{T^*}{T}\right) + d(\delta W)\right\} - T^*\delta S_p}$$

$$\varepsilon = H - T^*S, \quad \varepsilon_i = H_i - T^*S_i, \quad \varepsilon_j^* = 0$$

BE's per unit time

[U + X]: $$\frac{d}{dt}(U + X) = -\Delta(d\dot{H} + d\dot{X} + d\dot{Q} + d\dot{Q}^* + d\dot{W})$$

[S]: $$\frac{dS}{dt} = -\Delta\left(d\dot{S} + \frac{d\dot{Q}}{T} + \frac{d\dot{Q}^*}{T^*}\right) + \dot{S}_p$$

[U + X] − T*[S]:

$$\boxed{\frac{d}{dt}(U + X - T^*S) = -\Delta\left\{d\dot{\varepsilon} + d\dot{X} + d\dot{Q}\left(1 - \frac{T^*}{T}\right) + d\dot{W}\right\} - T^*\dot{S}_p}$$

2.2.4 Interface Transport and Equilibrium Conditions

We shall now apply the FBE's formulated in 2.2.1 to an interface system. The system considered is bounded by two planes in adjacent phases an infinitesimal distance apart from each other.
In the FBE's of this system the accumulation terms become zero. Zero production terms in the FBE's result in continuôus values of the transport of volume by displacement, δV_x, the transport of n_i, δn_i, and the transport of U, δU_x, when crossing the interface. The three contributions to δU_x, enthalpy transport by matter, heat and displacement work, do change mutually when crossing the interface. The transport δS_x is only continuous when all transport is reversible. It appears that the entropy producton δS_p equals a sum of products. Each product involves a transport of an extensive over a time interval and a driving force or drop in potential. As δS_p is always positive or zero, it follows that energy is transported in the direction of lower T, an interface is displaced in the direction of lower p, while n_i is transported in the direction of lower μ_i. Equating δS_p to zero yields the interface equilibrium conditions.

We now focus on special cases of interface transport. The transport of heat, δQ is in the direction of lower T. When T_1 is constant, the transport of heat continues until T_2 has become equal to T_1. A thermometer is based on this principle.

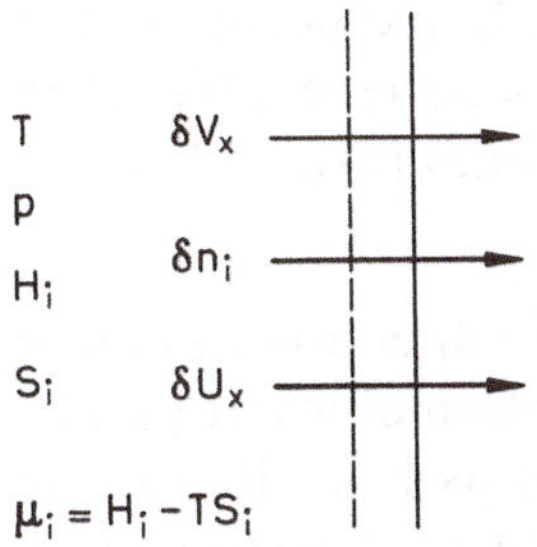

$$
\begin{array}{lll}
[V]: & 0 = -\Delta(\delta V_x) & \\
[n_i]: & 0 = -\Delta(\delta n_i) & \\
[U]: & 0 = -\Delta(\delta U_x) & \delta U_x = \sum H_i \delta n_i + \delta Q - p\delta V_x \\
[S]: & 0 = -\Delta(\delta S_x) + \delta S_p & \delta S_x = \sum S_i \delta n_i + \dfrac{\delta Q}{T} \\
 & & \quad = \delta U_x \dfrac{1}{T} + \delta V_x \dfrac{p}{T} + \sum \delta n_i \left(-\dfrac{\mu_i}{T}\right)
\end{array}
$$

δV_x, δn_i and δU_x are continuous

$$
\boxed{\delta S_p = \Delta(\delta S_x) = \delta U_x \Delta\left(\frac{1}{T}\right) + \delta V_x \Delta\left(\frac{p}{T}\right) + \sum \delta n_i \left(-\Delta \frac{\mu_i}{T}\right)}
$$

Interface equilibrium T, p and μ_i's are gradientless

Transport of heat

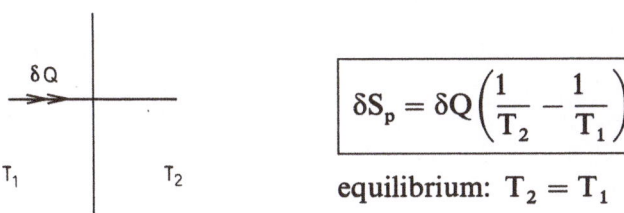

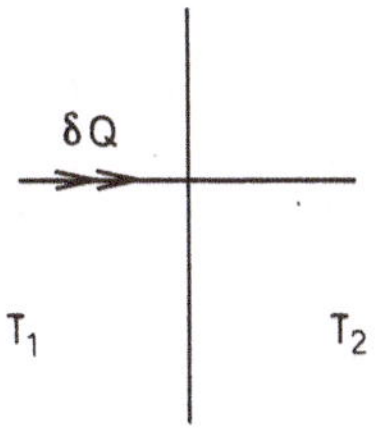

$$
\boxed{\delta S_p = \delta Q \left(\frac{1}{T_2} - \frac{1}{T_1}\right)}
$$

equilibrium: $T_2 = T_1$

Isothermal transport of n_i through a selective membrane occurs in the direction of lower μ_i. The membrane transport becomes reversible when $\Delta\mu_i = 0$. Non-isothermal membrane transport becomes reversible when $\Delta H_i = \Delta S_i = 0$.

Transport of volume by displacement, δV_x, is in the direction of higher p (displacement of the closed interface is in the direction of lower p). The net work by the phases on the interface is rejected as heat. Reversible displacement can be visualised as the displacement of a piston in an idealised piston-in-cylinder system. The net work on the interface is no longer rejected as heat, but constitutes a work output.

Isothermal membrane transport

irreversible

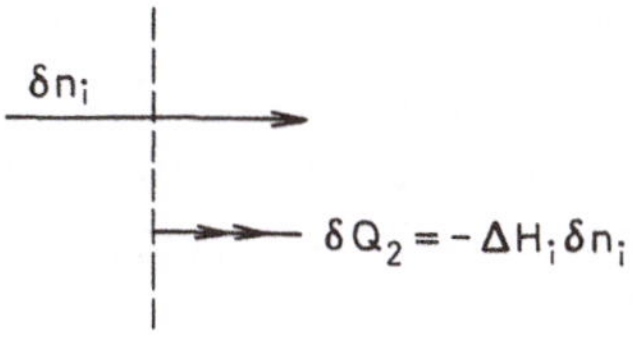

$$\delta S_p = -\,\delta n_i \frac{\Delta \mu_i}{T}$$

reversible

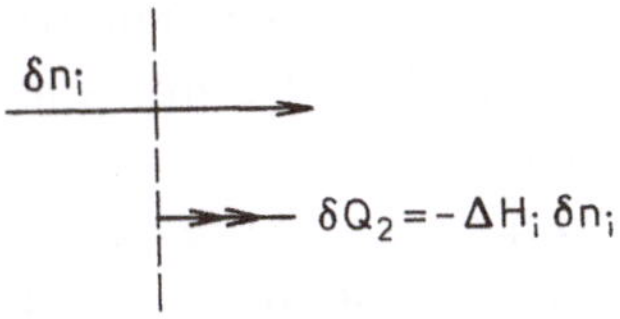

$$\Delta \mu_i = 0$$

Non-isothermal membrane transport

irreversible

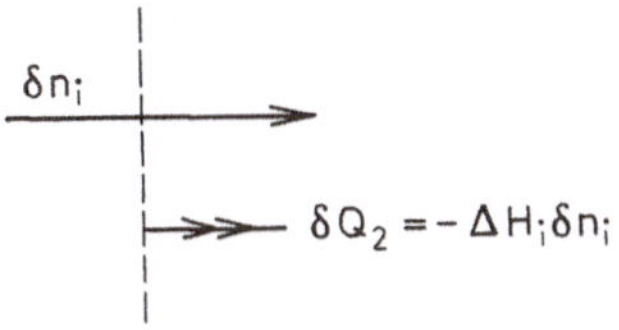

$$\delta S_p = \left(\Delta S_i - \frac{\Delta H_i}{T_2}\right)\delta n_i$$

reversible

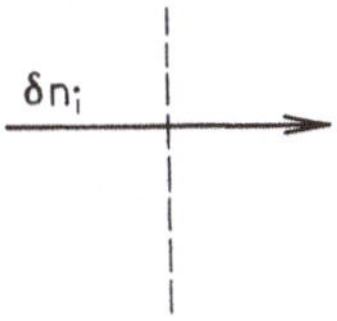

$$\Delta H_i = \Delta S_i = 0$$

Displacement

irreversible

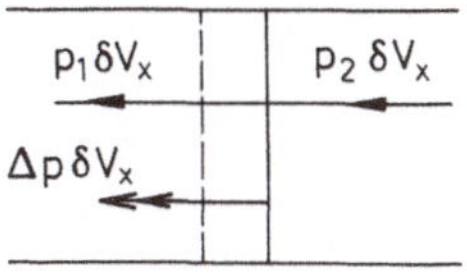

$$\delta S_p = \frac{\Delta p \, \delta V_x}{T_1}$$

reversible

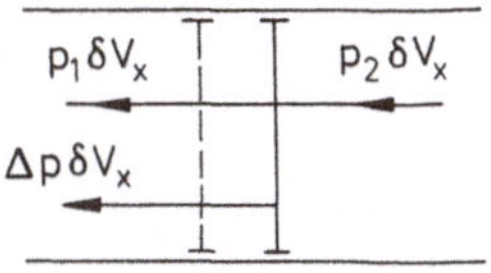

$$dW^{-} = \Delta p \, \delta V_x$$

2.2.5 FPR of Matter

We are now ready to derive the fundamental PR (FPR) of matter. We consider a multi-phase system with established interface equilibria and gradientless intensives in the individual phases. T, p and μ_i's are gradientless throughout the system. The transport between two phases is reversible: in addition to δV_x, δn_i and δU_x, the entropy transport δS_x is continuous. The transport of volume by displacement, the non-isothermal membrane transport and the transport of heat from the environment into each phase of the system are reversible as well.

Application of the interface equilibrium conditions and the BE's [V], [n_i], [U] and [S] of the system yields the FPR of matter. It shows that U, S, V and n_i's of an equilibrium state are interrelated. The FPR can be interpreted as a condition for intraphase physical equilibrium.

When the transport of matter into the system is zero (closed system), the reversible transport of heat, dQ is readily related to changes of system properties. The FPR of a closed system involves U, S and V.

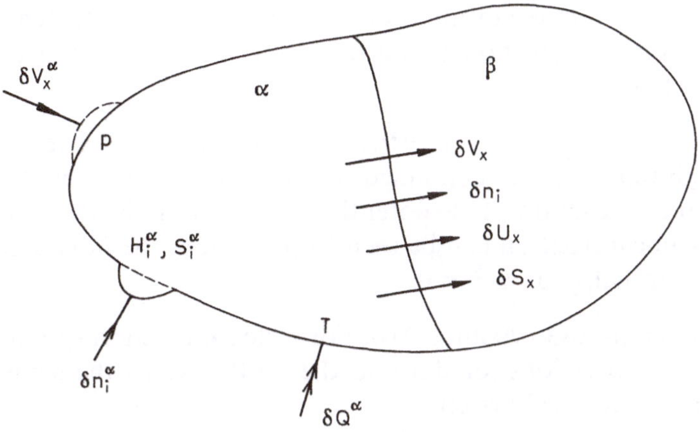

Interface equilibrium $\quad T^\alpha = T^\beta = T, \quad p^\alpha = p^\beta = p, \quad \mu_i^\alpha = \mu_i^\beta = \mu_i$

[V]: $$dV = \sum_\alpha \delta V_x^\alpha$$

[n_i]: $$dn_i = \sum_\alpha \delta n_i^\alpha$$

[U]: $$dU = \sum_\alpha \delta U_x^\alpha = \sum_\alpha \sum_i H_i^\alpha \delta n_i^\alpha + \sum_\alpha dQ^\alpha - \sum_\alpha p \delta V_x^\alpha$$
$$= \sum_\alpha \sum_i H_i^\alpha \delta n_i^\alpha + dQ - pdV$$

[S]: $$dS = \sum_\alpha \delta S_x^\alpha = \sum_\alpha \sum_i S_i^\alpha \delta n_i^\alpha + \sum_\alpha \frac{dQ^\alpha}{T} = \sum_\alpha \sum_i S_i^\alpha \delta n_i^\alpha + \frac{dQ}{T}$$

[U] − T[S]: $$dU - TdS = \sum_\alpha \sum_i (H_i^\alpha - TS_i^\alpha)\delta n_i^\alpha - pdV$$
$$= \sum_\alpha \sum_i \mu_i \delta n_i^\alpha - pdV = \sum_i \mu_i dn_i - pdV$$

FPR: $$\boxed{dU = TdS - pdV + \sum \mu_i dn_i}$$

Closed system

$$\boxed{dQ = dU + pdV = TdS}$$

$$\boxed{dU = TdS - pdV}$$

2.2.6 Chemical Equilibrium Condition

To arrive at the equilibrium condition for a chemical reaction we extend the system of the previous section by allowing a chemical reaction to proceed in the system.

The BE's [n_i], [U] and [S], and the FPR which reflects that physical equilibrium is maintained can be combined to give an expression for δS_p due to the chemical reaction. The potential for reaction is the free enthalpy. The chemical reaction can be seen to proceed in the direction of decreasing free enthalpy or $\Delta\bar{G} < 0$.

For a reversible reaction $\Delta\bar{G}$ becomes zero. For a closed reaction system at equilibrium, the expressions for dQ and the FPR assume the same forms as for a closed physical system.

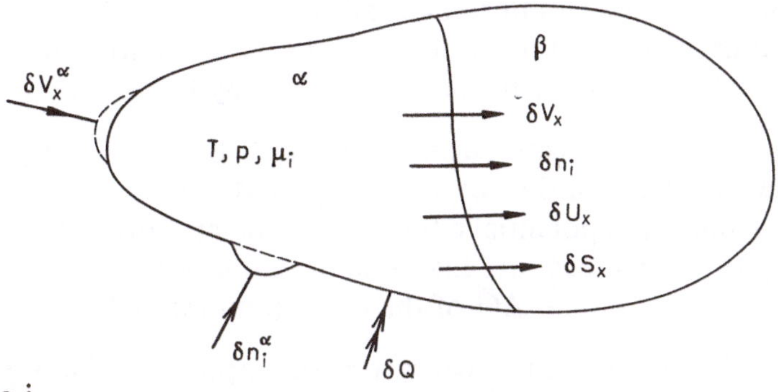

reaction:

$0 \rightarrow \sum \nu_i$ mole i

$[n_i]$: $\quad dn_i = \sum_\alpha \delta n_i^\alpha + \nu_i \delta n_{kp}$

$[U]$: $\quad dU = \sum_\alpha \delta U_x^\alpha = \sum_\alpha \sum_i H_i^\alpha \delta n_i^\alpha + \delta Q - pdV$

$[S]$: $\quad dS = \sum_\alpha \delta S_x^\alpha = \sum_\alpha \sum_i S_i^\alpha \delta n_i^\alpha + \frac{\delta Q}{T} + \delta S_p$

FPR: $\quad dU = TdS - pdV + \sum \mu_i dn_i$

$[U] - T[S] - \sum \mu_i [n_i]$:

$$-pdV = -pdV - T\delta S_p - \sum \nu_i \mu_i \delta n_{kp}$$

$$\boxed{\delta S_p = \frac{-\sum \nu_i \mu_i \delta n_{kp}}{T} = -\frac{\Delta \bar{G}\, \delta n_{kp}}{T}}$$

chemical equilibrium

$$\boxed{\sum \nu_i \mu_i = \Delta \bar{G} = 0}$$

Closed system

$$\boxed{dQ = dU + pdV = TdS}$$

$$\boxed{dU = TdS - pdV}$$

2.2.7 Electrochemical Equilibrium Condition

To arrive at the equilibrium condition for an electrochemical reaction the describing equations in the previous section are extended to include electrochemical work in the internal energy balance [U].

For a reversible electrochemical reaction the electrical potential ϕ assumes the value of the equilibrium potential e.
The equilibrium potential e turns out to be proportional to $-\Delta\bar{G}$ of the electrochemical reaction. Likewise the equilibrium electrode potential e_i is proportional to $-\Delta\bar{G}$ of the corresponding redox reaction.

For a closed electrochemical system the expressions for dQ and the FPR include a term representing reversible electrochemical work in addition to the term for reversible expansion work.

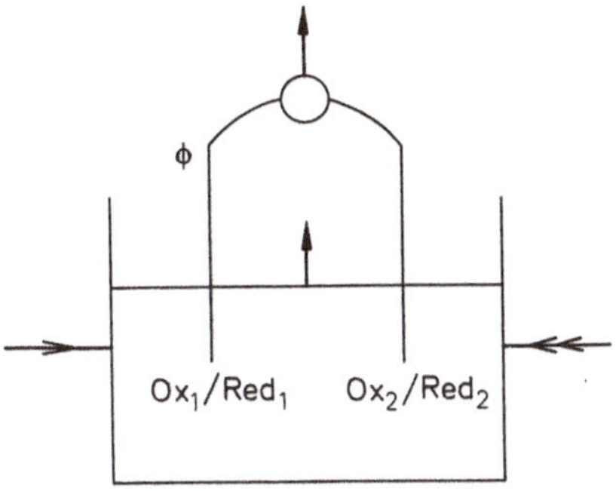

$$\nu_{Ox_1} Ox_1 \rightarrow \nu_{Red_1} Red_1 + 1 \oplus$$
$$\nu_{Red_2} Red_2 + 1 \oplus \rightarrow \nu_{Ox_2} Ox_2$$

$[n_i]$: $\quad dn_i = \sum \delta n_i^\alpha + \nu_i \delta n_{kp}$

$[U]$: $\quad dU = \sum \delta U_x^\alpha = \sum_\alpha \sum_i H_i^\alpha \delta n_i^\alpha + \delta Q - pdV - \phi \mathscr{F} \delta n_{kp}$

$[S]$: $\quad dS = \sum \delta S_x^\alpha = \sum_\alpha \sum_i S_i^\alpha \delta n_i^\alpha + \dfrac{\delta Q}{T} + \delta S_p$

FPR: $\quad dU = TdS - pdV + \sum \mu_i dn_i$

$[U] - T[S] - \sum \mu_i [n_i]$:

$$-pdV = -pdV - \phi \mathscr{F} \delta n_{kp} - \sum \nu_i \mu_i \delta n_{kp} - T\delta S_p$$

equilibrium potential

$$e = \frac{-\sum \nu_i \mu_i}{\mathscr{F}} = \frac{-\Delta \bar{G}}{\mathscr{F}}$$

$$\delta S_p = \frac{(e - \phi) \mathscr{F} \delta n_{kp}}{T}$$

electrode potential

$$e_i = \frac{-\Delta \bar{G}_{Ox_i/Red_i}}{\mathscr{F}} = \frac{\nu_{Ox_i} \mu_{Ox_i} - \nu_{Red_i} \mu_{Red_i}}{\mathscr{F}}$$

$$e = e_1 - e_2$$

Closed system

$$dQ = dU + pdV + e\mathscr{F} dn_k = TdS$$

$$dU = TdS - pdV - e\mathscr{F} dn_k$$

3 PR's of Single-Phase Systems

After the principles postulated in Chapter 1 and the first deductions presented in Chapter 2, we shall now derive PR's of a single-phase system in the three consecutive cells of Chapter 3. Inputs to each cell can come from cells with equal or lower row and column numbers.

3.1 Column-1 PR's of Single-Phase Systems

In this cell we consider a single-phase system of N components. An extensive equilibrium state of such a system has $2 + N$ independent properties, whereas an intensive state has $1 + N$ degrees of freedom, for example T, p and $N - 1$ independent mole fractions x_i. Starting points in deriving column-1 PR's are the extensive property differential $dE(T, p, n_i)$ and the addition rule presented in cell 1.1.

First we derive the Gibbs–Duhem relation for a generalised extensive E. This PR between the $2 + N$ primary intensive properties, T, p and the N E_i's, reduces the number of independent intensives to $1 + N$.

Further the intensive property differentials $d\bar{E}(T, p, x_i)$ and $dE_i(T, p, x_j)$, and the differentiation rule are derived as column-1 PR's. With the differentiation rule the partial molar properties E_i can be found when $\bar{E}$ is known as a function of composition.

Finally, we define, for the sake of convenience, a number of generalised properties: the property of mixing ΔE_M, the ideal-state property E^{is}, the non-ideality property E^{ni} as well as the standard property $E_i^0(T)$, the standard property of formation $\Delta E_{f,i}^0$ and the standard property of reaction ΔE^0.

3.2 Column-2 PR's of Single-Phase Systems

Main inputs into this cell are in addition to cell 3.1, the FPR of matter and the FBE's for reversible heating of a closed system as derived in Sect. 2.2.5. Further we define the heat capacities C_p and C_v. Outputs are PR's which interconnect functions of state, express functions of state in terms of measurable properties or interrelate measurables.

Equivalent representations of the FPR $dU(S, V, n_i)$ are readily found from the definitions of the auxiliary energy functions. They are the total property differentials $dH(S, p, n_i)$, $dF(T, V, n_i)$ and $dG(T, p, n_i)$.

Cross-differentiation identities of $dF(T, V, n_i)$ and $dG(T, p, n_i)$ provide examples of column-2 PR's which express functions of state in terms of measurable properties. Combined with experimental knowledge of T-p-V behaviour they reveal that S increases with increasing V and decreasing p. This is consistent with the microscopic concept that entropy is a measure of the disorder of molecules: the further away from an ordered crystalline structure, the higher the entropy.

Examples of PR's which interconnect functions of state are again cross-differentiation identities of $dG(T, p, n_i)$. They make it possible to find the partial molar properties S_i, V_i and H_i once $\mu_i(T, p, x_i)$ is known.

The total property differentials for a closed system give rise to several column-2 PR's between measurable properties. An example is the expression for $\bar{C}_p - \bar{C}_V$ which on the basis of experimental T-p-$\bar{V}$ behaviour is a positive quantity.

There are many alternative representations of the FPR and its equivalents and PR's of the type $dE(T, p, n_i)$. They involve properties of mixing, ideal-state and non-ideality properties, average and partial molar properties as well as average and partial specific properties. They are summarised by the set of single-phase properties presented in Sect. 3.2.3.

3.3 Column-3 PR's of Single-Phase Systems

Here we bring into play the phenomenological laws on asymptotic phase behaviour postulated in cell 1.3. They are PR's which all liquids or gases have in common and express the chemical potentials μ_i in terms of measurable properties. Further we use as an approximative experimental law the expression

$$C^0_{p,i}(T) = A_i + B_i T$$

for the standard heat capacity of a component and apply as approximations zero thermal expansion and compressibility for ideal-state liquids and solids. The resultant PR's are obtained by employing the column-2 PR's presented in cell 3.2, particularly the set of single-phase PR's.

First we derive PR's for the standard properties of heating for G_i/T or μ_i/T, H_i, S_i and ε_i. They are equations to calculate standard properties at a temperature T from tabulated values at a reference temperature T_0.

We define as ideal states the infinite-dilution, the zero-dilution and the ideal-gas states, as corresponding non-ideality properties the excess properties $E^{e,\infty}$ and $E^{e,z}$ and the residual property E^r, and as non-ideality coefficients the activity coefficients γ_i^∞ and γ_i^z, and the fugacity coefficient φ_i. By definition the non-ideality properties vanish and the non-ideality coefficients become unity at the conditions where the laws on asymptotic phase behaviour apply. On this basis μ_i-models are devised for liquids and gases.

PR's for the important partial molar properties of solids, liquids and gases are readily found. They have as increments ideal-state properties of compression and mixing, and non-ideality properties. The equations serve to calculate real-state properties from the standard property at the same T.

Special attention is paid to derive PR's to calculate the partial molar exergy of a compound in a stream as a function of T, p and composition for a given infinite reservoir with temperature T*, pressure p* and a set of components j having zero partial molar exergies.
A first set of PR's gives the standard exergies $\varepsilon_1^0(T^*)$ of the elements to match the conditions of the infinite reservoir using tabulated standard properties as input. Having the standard exergies of the elements, a second set of PR's calculates the desired value of the partial molar exergy of the compound from the standard exergies of its constituting elements and tabulated standard properties of the compound.

This cell is concluded with the development of PR's for a closed ideal-gas system, PR's which express residual properties of a gas in terms of measurables, and, finally, definitions and PR's involving activity coefficients of a solvent and its solutes.

3.1.1 Extensive State

In Sect. 1.1.2 we postulated that any extensive E of an equilibrium state of a system is a function of 2 + N independent properties, e.g. T, p and N n_i's or m_i's, and independent of the path by which the equilibrium state is reached. This led to the total property differential dE and the definitions of the intensives E_i and e_i associated with E. Further we integrated dE at constant intensives to obtain the addition rule.

Here we combine dE and the addition rule to find the Gibbs–Duhem relation. This PR reflects that at given T and p only N − 1 of the N E_i's of an extensive E are independent.

The second derivatives of E do not depend on the order of differentiation. This results in cross-differentiation identities of the total differential dE. The derivatives of E_i and e_i w.r.t. T and p will be applied in the next section.

Finally, we define some extensives which can be substituted for E in the PR's presented. The property of mixing ΔE_M is an extensive property of a mixture relative to its constituent pure components at the same T and p. The split of an extensive in an ideal-state part E^{is} and a non-ideality part E^{ni} will turn out to be convenient when applying the phenomenological laws on asymptotic phase behaviour.

Total property differential

$$dE = E_T\,dT + E_p\,dp + \sum E_i\,dn_i = E_T\,dT + E_p\,dp + \sum e_i\,dm_i$$

first derivatives:

$$E_T = \frac{\partial E}{\partial T}, \quad E_p = \frac{\partial E}{\partial p}, \quad E_i = \frac{\partial E}{\partial n_i}, \quad e_i = \frac{\partial E}{\partial m_i} = \frac{E_i}{M_i}$$

second derivatives (cross-differentiation identities):

$$\frac{\partial E_T}{\partial p} = \frac{\partial E_p}{\partial T}, \quad \frac{\partial E_i}{\partial T} = E_{T,i}, \quad \frac{\partial E_i}{\partial p} = E_{p,i}$$

$$\frac{\partial e_i}{\partial T} = e_{T,i}, \quad \frac{\partial e_i}{\partial p} = e_{p,i}$$

Addition rule

$$E = \sum n_i E_i = \sum m_i e_i$$

Gibbs–Duhem relation

$$\sum_i n_i\,dE_i = \sum_i m_i\,de_i = E_T\,dT + E_p\,dp$$

$$\text{or} \quad \sum n_i (dE_i)_{T,p} = 0$$

Substitute extensives

property of mixing: $\Delta E_M = E - \sum n_i \bar{E}_i$

$$\bar{E}_i \equiv \bar{E}(x_i = 1) = E_i(x_i = 1)$$

ideal-state properties: E^{is}

non-ideality properties: $E^{ni} = E - E^{is}$

3.1.2 Intensive State

We postulated that the extensive and intensive states of a single-phase system are specified by a set of $2 + N$ independent properties. Possible sets are T, p and N n_i's, and T, p, $N - 1$ independent x_i's and n. The set must contain at least one extensive property.
The intensive state alone is specified by $1 + N$ independent intensives, e.g. T, p and $N - 1$ independent x_i's
Alternatively we could select T, p and N E_i's as primary variables. The intensive-state version of the Gibbs–Duhem relation between these variables reduces the number of independent intensives again to $1 + N$.

The intensive property differential $d\bar{E}$ is found by combining the addition rule and the Gibbs–Duhem relation. The total differential dE_i follows by inserting the derivatives of E_i w.r.t. T and p as found in the previous section.

$\bar{E}$ can be found from the E_i's by using the addition rule. How can, vice versa, the E_i's be found when $\bar{E}$ is known as a function of composition? The differentiation rule serves this purpose. It is derived from the derivative of $\bar{E}$ w.r.t. x_i which can be read from the total differential $d\bar{E}$. For a binary system E_1 and E_2 can be obtained from a plot of $\bar{E}$ versus x_2.

Addition rule

$$\bar{E} = \sum x_i E_i$$

Gibbs–Duhem relation

$$\sum_i x_i \, dE_i = \bar{E}_T \, dT + \bar{E}_p \, dp$$

$$\text{or } \sum_i x_i (dE_i)_{T,p} = 0, \quad \sum_i x_i \frac{\partial E_i}{\partial x_{j \neq k}} = 0$$

$$\text{binary: } x_1 \frac{\partial E_1}{\partial x_2} + x_2 \frac{\partial E_2}{\partial x_2} = 0$$

Differential of average molar property

$$d\bar{E} = \bar{E}_T \, dT + \bar{E}_p \, dp + \sum_i E_i \, dx_i$$
$$= \bar{E}_T \, dT + \bar{E}_p \, dp + \sum_{i \neq k} (E_i - E_k) \, dx_i$$

Differential of partial molar property

$$dE_i = E_{T,i} dT + E_{p,i} dp + \sum_{j \neq k} \frac{\partial E_i}{\partial x_j} dx_j$$

Differentiation rule

$$\frac{\partial \bar{E}}{\partial x_{i \neq k}} = E_i - E_k$$

$$E_k = \bar{E} - \sum_{i \neq k} x_i \frac{\partial \bar{E}}{\partial x_i}$$

$$\text{binary: } E_1 = \bar{E} - x_2 \frac{\partial \bar{E}}{\partial x_2}$$

$$E_2 = \bar{E} - x_1 \frac{\partial \bar{E}}{\partial x_1} = \bar{E} + (1 - x_2) \frac{\partial \bar{E}}{\partial x_2}$$

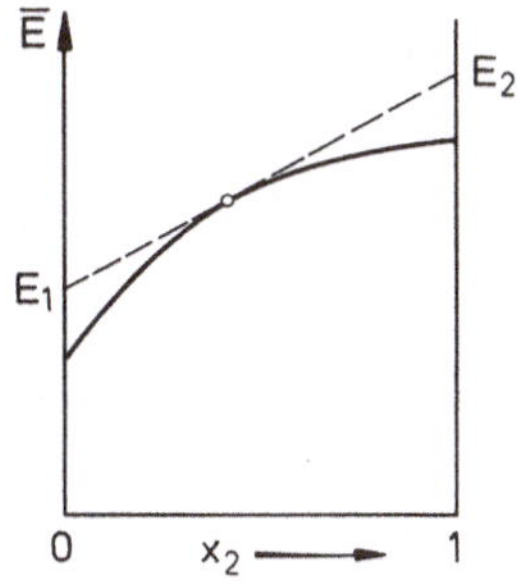

For the sake of convenience we define the standard property $E_i^0(T)$. It is the molar pure-component property in the ideal state at standard pressure p_0.
The standard property of formation $\Delta E_{f,i}^0$ is the standard property of a compound i relative to the sum of the standard properties of its constituent elements. It is zero for an element.
The standard property of a reaction, ΔE^0 is the change in standard property accompanying that reaction. It can be found from the values of $\Delta E_{f,i}^0$ for the reaction components.

So far we described the intensive state of a single-phase system by PR's on a mole basis. An alternative description is given by the analogous PR's on a mass basis.

Standard properties

standard property: $$\boxed{E_i^0(T) \equiv \bar{E}_i^{is}(T, p_0)}$$

standard property of formation: $\sum_l b_{li}\,\text{mole } l \rightarrow 1 \text{ mole } i$

$$\boxed{\Delta E_{f,i}^0 \equiv E_i^0 - \sum_l b_{li} E_l^0} \qquad M_i = \sum b_{li} M_l$$

standard property of reaction: $0 \rightarrow \sum v_i \,\text{mole } i$

$$\sum_i v_i M_i = 0, \quad \sum v_i b_{li} = 0$$

$$\boxed{\Delta E^0 \equiv \sum_i v_i E_i^0 = \sum_i v_i \Delta E_{f,i}^0}$$

Intensive-state PR's on a mass basis

addition rule: $\bar{e} = \sum w_i e_i$

Gibbs–Duhem relation: $\sum w_i\, de_i = \bar{e}_T\, dT + \bar{e}_p\, dp$

$$\sum_i w_i (de_i)_{T,p} = 0, \quad \sum_i w_i \frac{\partial e_i}{\partial x_{j \neq k}} = 0$$

differential of average specific prop.

$$d\bar{e} = \bar{e}_T\, dT + \bar{e}_p\, dp + \sum_i e_i\, dw_i$$
$$= \bar{e}_T\, dT + \bar{e}_p\, dp + \sum_{i \neq k} (e_i - e_k)\, dw_i$$

differentiation rule: $$e_k = \bar{e} - \sum_{i \neq k} w_i \frac{\partial \bar{e}}{\partial w_i}$$

3.2.1 FPR and Equivalents. Reversible Heating. Functions of T, V, n_i and T, p, n_i

Having completed the derivation of generalised PR's, we have now arrived at the column-2 part of Chapter 3. In this part the FPR of matter and the expressions for the reversible heating of a closed system will be developed into single-phase PR's. PR's which interrelate functions of state, PR's which relate functions of state to measurable properties as well as PR's between measurables will emerge.

Using the definitions of H, F and G three alternative representations of the FPR are readily found. Each representation has its characteristic set of independent properties. Of the alternative representations $dF(T, V, n_i)$ and especially $dG(T, p, n_i)$ will turn out to be useful as their independents are all measurable. Knowledge of the functions $F(T, V, n_i)$ or $G(T, p, n_i)$ suffices to find all other functions by differentiation and application of definitions. To G a Gibbs–Duhem relation can be attached.

The four Maxwell relations are found as cross-differentiation identities at constant n_i's. The ones associated with dF and dG connect S with measurables. From experimental knowledge on T-p-V behaviour it follows that S increases with increasing V and decreasing p. This is consistent with the concept that S is a measure of the disorder of matter: the further away from an ordered crystalline structure, the higher the entropy.

Of the other cross-differentiation identities indicated the one associated with dF connects μ_i with measurables. When the function $\mu_i(T, p, x_i)$ is known, all other partial molars can be found from the cross-differentiation identities of dG and their definitions.

To the expressions for the reversible heating of a closed system we add the definitions of the heat capacities at constant V and at constant p, the extensives C_v and C_p.

The single-phase PR's developed above suffice to find the total differentials of $S(T, V, n_i)$, $U(T, V, n_i)$, $S(T, p, n_i)$ and $H(T, p, n_i)$.

energy function	FPR and equivalents	Maxwell rel. n_i = constant	other cross-diff. identies
$U(S, V, n_i)$	$dU = \quad T\,dS - p\,dV + \sum \mu_i\,dn_i$	$\left(\frac{\partial T}{\partial V}\right)_S = -\left(\frac{\partial p}{\partial S}\right)_V$	
$H = U + pV$ $H(S, p, n_i)$	$dH = \quad T\,dS + V\,dp + \sum \mu_i\,dn_i$	$\left(\frac{\partial T}{\partial p}\right)_S = \left(\frac{\partial V}{\partial S}\right)_p$	
$F = U - TS$ $F(T, V, n_i)$	$dF = -S\,dT - p\,dV + \sum \mu_i\,dn_i$	$\left(\frac{\partial S}{\partial V}\right)_T = \left(\frac{\partial p}{\partial T}\right)_V$	$\frac{\partial \mu_i}{\partial V} = -\frac{\partial p}{\partial n_i}$
$G = U + pV - TS$ $G(T, p, n_i)$	$dG = -S\,dT + V\,dp + \sum \mu_i\,dn_i$	$\left(\frac{\partial S}{\partial p}\right)_T = -\left(\frac{\partial V}{\partial T}\right)_p$	$S_i = -\frac{\partial \mu_i}{\partial T}$
	Gibbs–Duhem relation:		$V_i = \frac{\partial \mu_i}{\partial p}$
	$\sum x_i\,d\mu_i = -\bar{S}\,dT + \bar{V}\,dp$		$H_i = -T^2 \frac{\partial}{\partial T}\left(\frac{\mu_i}{T}\right)$

Reversible heating of a closed system

$$dQ = dU + p\,dV = T\,dS, \quad dU = T\,dS - p\,dV$$

heat capacities: $C_V \equiv \left(\frac{dQ}{dT}\right)_V, \quad C_p \equiv \left(\frac{dQ}{dT}\right)_p$

Functions of T, V, n_i and of T, p, n_i

$S(T, V, n_i)$	$dS = \frac{C_v}{T}\,dT + \frac{\partial p}{\partial T}\,dV + \sum \frac{\partial S}{\partial n_i}\,dn_i$
$U(T, V, n_i)$	$dU = C_v\,dT + \left(T\frac{\partial p}{\partial T} - p\right)dV + \sum \frac{\partial U}{\partial n_i}\,dn_i$
$S(T, p, n_i)$	$dS = \frac{C_p}{T}\,dT - \frac{\partial V}{\partial T}\,dp + \sum S_i\,dn_i$
$H(T, p, n_i)$	$dH = C_p\,dT + \left(V - T\frac{\partial V}{\partial T}\right)dp + \sum H_i\,dn_i$

3.2.2 Closed System

From the PR's generated in the previous section the total differentials of the functions $\bar{S}(T, p)$, $\bar{H}(T, p)$, $\bar{S}(T, \bar{V})$, $\bar{U}(T, \bar{V})$, and $\bar{S}(p, \bar{V})$ of a closed system can be formulated.

Additional measurable properties defined are the thermal expansion coefficient α, the isothermal compressibility K_T and the insentropic compressibility K_S.

A number of PR's between measurables present themselves: the change of $\bar{C}_p$ with p as found from a cross-differentiation identity, the Joule–Thomson effect, the change of $\bar{C}_v$ with $\bar{V}$, the difference $\bar{C}_p - \bar{C}_v$ which on the basis of experimental T-p-$\bar{V}$ behaviour is positive, and the relation between K_S and K_T of which the isentropic compressibility is the smaller one.

Functions of measurables	PR's between measurables
$d\bar{V} = \frac{\partial \bar{V}}{\partial T} dT + \frac{\partial \bar{V}}{\partial p} dp$ $d\bar{S} = \frac{\bar{C}_p}{T} dT - \frac{\partial \bar{V}}{\partial T} dp$ $d\bar{H} = \bar{C}_p dT + \left(\bar{V} - T\frac{\partial \bar{V}}{\partial T}\right) dp$	$\alpha \equiv \frac{1}{\bar{V}}\left(\frac{\partial \bar{V}}{\partial T}\right)_p \quad K_T \equiv -\frac{1}{\bar{V}}\left(\frac{\partial \bar{V}}{\partial p}\right)_T$ $\frac{1}{T}\frac{\partial \bar{C}_p}{\partial p} = -\frac{\partial^2 \bar{V}}{\partial T^2}$ $\boxed{\left(\frac{\partial T}{\partial p}\right)_{\bar{H}} = \frac{T\frac{\partial \bar{V}}{\partial T} - \bar{V}}{\bar{C}_p}}$ (Joule–Thomson)
$dp = \frac{\partial p}{\partial T} dT + \frac{\partial p}{\partial \bar{V}} d\bar{V}$ $d\bar{S} = \frac{\bar{C}_v}{T} dT + \frac{\partial p}{\partial T} d\bar{V}$ $d\bar{U} = \bar{C}_v dT + \left(T\frac{\partial p}{\partial T} - p\right) d\bar{V}$	$\frac{1}{T}\frac{\partial \bar{C}_v}{\partial \bar{V}} = \frac{\partial^2 p}{\partial T^2}$ $\bar{C}_p - \bar{C}_v = T\left(\frac{\partial \bar{S}}{\partial T}\right)_p - \bar{C}_v$ $= T\left\{\frac{\bar{C}_v}{T} + \left(\frac{\partial p}{\partial T}\right)_{\bar{V}}\left(\frac{\partial \bar{V}}{\partial T}\right)_p\right\} - \bar{C}_v$ $\boxed{\bar{C}_p - \bar{C}_v = T\left(\frac{\partial p}{\partial T}\right)_{\bar{V}}\left(\frac{\partial \bar{V}}{\partial T}\right)_p}$ $\boxed{\left(\frac{\partial T}{\partial \bar{V}}\right)_{\bar{U}} = \frac{p - T\frac{\partial p}{\partial T}}{\bar{C}_v}}$ (Joule)
$dT = \frac{\partial T}{\partial p} dp + \frac{\partial T}{\partial \bar{V}} d\bar{V}$ $d\bar{S} = \frac{\bar{C}_v}{T}\frac{\partial T}{\partial p} dp + \frac{\bar{C}_p}{T}\frac{\partial T}{\partial \bar{V}} d\bar{V}$	$\left(\frac{\partial \bar{V}}{\partial p}\right)_T = -\left(\frac{\partial T}{\partial p}\right)_{\bar{V}}\left(\frac{\partial \bar{V}}{\partial T}\right)_p$ $\left(\frac{\partial \bar{V}}{\partial p}\right)_{\bar{S}} = \frac{\bar{C}_v}{\bar{C}_p}\left(\frac{\partial \bar{V}}{\partial p}\right)_T$ $K_S \equiv -\frac{1}{\bar{V}}\left(\frac{\partial \bar{V}}{\partial p}\right)_{\bar{S}}$ $\boxed{K_S = \frac{\bar{C}_v}{\bar{C}_p} K_T}$

3.2.3 Set a Single-Phase PR's

We shall conclude the column-2 part of this chapter with the many alternative representations of the FPR and its equivalents and of PR's of the type $dE(T, p, n_i)$. They are obtained by replacing the extensives A, $\Sigma\mu_i\, dn_i$ and $\Sigma E_i\, dn_i$ occurring in the set of single-phase PR's, by one of the substitute sets. The resultant PR's involve properties of mixing, ideal-state and non-ideality properties, average and partial molars as well as average and partial specific properties.

FPR and equivalents

U	$dU = T\,dS - p\,dV + \sum \mu_i\,dn_i$	
$H = U + pV$	$dH = T\,dS + V\,dp + \sum \mu_i\,dn_i$	
$F = U - TS$	$dF = -S\,dT - p\,dV + \sum \mu_i\,dn_i$	
$G = U + pV - TS$	$dG = -S\,dT + V\,dp + \sum \mu_i\,dn_i$	$S = -\frac{\partial G}{\partial T}$, $V = \frac{\partial G}{\partial p}$, $H = -T^2 \frac{\partial}{\partial T}\left(\frac{G}{T}\right)$

Property differentials $dE(T, p, n_i)$

E	$dE = E_T\,dT + E_p\,dp + \sum E_i\,dn_i$
S	$dS = \frac{C_p}{T}\,dT - \frac{\partial V}{\partial T}\,dp + \sum S_i\,dn_i$
H	$dH = C_p\,dT + \left(V - T\frac{\partial V}{\partial T}\right)dp + \sum H_i\,dn_i$
$\frac{G}{T}$	$d\frac{G}{T} = -\frac{H}{T^2}\,dT + \frac{V}{T}\,dp + \sum \frac{\mu_i}{T}\,dn_i$
$\varepsilon = H - T^*S$	$d\varepsilon = C_p\left(1 - \frac{T^*}{T}\right)dT + \left\{V - (T - T^*)\frac{\partial V}{\partial T}\right\}dp + \sum \varepsilon_i\,dn_i$

Substitute sets of extensives A, $\sum \mu_i\,dn_i$ and $\sum E_i\,dn_i$

A, $\sum \mu_i\,dn_i$, $\sum E_i\,dn_i$	$\bar{A}$, $\sum \mu_i\,dx_i$, $\sum E_i\,dx_i$	A_i, $(d\mu_i)_{T,p}$, $(dE_i)_{T,p}$
ΔA_M, $\sum \Delta G_{M,i}\,dn_i$, $\sum \Delta E_{M,i}\,dn_i$	$\Delta\bar{A}_M$, $\sum \Delta G_{M,i}\,dx_i$, $\sum \Delta E_{M,i}\,dx_i$	$\Delta A_{M,i}$, $(d\Delta G_{M,i})_{T,p}$, $(d\Delta E_{M,i})_{T,p}$
A^{is}, $\sum \mu_i^{is}\,dn_i$, $\sum E_i^{is}\,dn_i$	$\bar{A}^{is}$, $\sum \mu_i^{is}\,dx_i$, $\sum E_i^{is}\,dx_i$	A_i^{is}, $(d\mu_i^{is})_{T,p}$, $(dE_i^{is})_{T,p}$
A^{ni}, $\sum G_i^{ni}\,dn_i$, $\sum E_i^{ni}\,dn_i$	$\bar{A}^{ni}$, $\sum G_i^{ni}\,dx_i$, $\sum E_i^{ni}\,dx_i$	A_i^{ni}, $(dG_i^{ni})_{T,p}$, $(dE_i^{ni})_{T,p}$
A, $\sum g_i\,dm_i$, $\sum e_i\,dm_i$	$\bar{a}$, $\sum g_i\,dw_i$, $\sum e_i\,dw_i$	a_i, $(dg_i)_{T,p}$, $(de_i)_{T,p}$

Examples of alternative representations read from the set of single-phase PR's elucidate the procedure.

The validity of the procedure for the PR's of the type $dE(T, p, n_i)$ including $dG(T, p, n_i)$ follow directly from cell 3.1. The alternative representations of dF, dH and dU are found from each alternative representation of dG using the definitions of the corresponding substitutes of F, H and U.

Examples

$$S_i = -\frac{\partial \mu_i}{\partial T}, \quad V_i = \frac{\partial \mu_i}{\partial P}, \quad H_i = -T^2 \frac{\partial}{\partial T}\left(\frac{\mu_i}{T}\right)$$

$$\Delta S_{M,i} = -\frac{\partial}{\partial T}\Delta G_{M,i}, \quad \Delta V_{M,i} = \frac{\partial}{\partial p}\Delta G_{M,i}, \quad \Delta H_{M,i} = -T^2 \frac{\partial}{\partial T}\left(\frac{\Delta G_{M,i}}{T}\right)$$

$$d\bar{H} = Td\bar{S} + \bar{V}\,dp + \sum \mu_i\,dx_i$$

$$dH_i = TdS_i + V_i\,dp + (d\mu_i)_{T,p}$$

$$d\bar{h} = Td\bar{s} + \bar{v}\,dp + \sum g_i\,dw_i$$

$$dh_i = Tds_i + v_i\,dp + (dg_i)_{T,p}$$

$$d\bar{H} = \bar{C}_p\,dT + \left(\bar{V} - T\frac{\partial \bar{V}}{\partial T}\right)dp + \sum H_i\,dx_i$$

$$dH_i = C_{p,i}\,dT + \left(V_i - T\frac{\partial V_i}{\partial T}\right)dp + (dH_i)_{T,p}$$

$$d\bar{h} = \bar{c}_p\,dT + \left(\bar{v} - T\frac{\partial \bar{v}}{\partial T}\right)dp + \sum h_i\,dw_i$$

$$dh_i = c_{p,i}\,dT + \left(v_i - T\frac{\partial v_i}{\partial T}\right)dp + (dh_i)_{T,p}$$

$$d\frac{\mu_i}{T} = -\frac{H_i}{T^2}\,dT + \frac{V_i}{T}dp + \left(d\frac{\mu_i}{T}\right)_{T,p}$$

$$d\varepsilon_i = C_{p,i}\left(1 - \frac{T^*}{T}\right)dT + \left\{V_i - (T - T^*)\frac{\partial V_i}{\partial T}\right\}dp + (d\varepsilon_i)_{T,p}$$

3.3.1 Standard Properties of Heating

Having explored the consequences of the FPR of matter and the expressions for reversible heating of a closed system, we have now come to the column-3 part of this chapter where we shall take into account the phenomenological laws on asymptotic phase behaviour for μ_i.

Starting from a generalised standard property $E_i^0(T_0) \equiv \bar{E}_i^{is}(T_0, p_0)$ at reference temperature T_0, a number of incremental properties need to be known to arrive at the real-state property $E_i(T, p, x_i)$. The standard property of heating leads to $E_i^0(T)$. Subsequently ideal-state properties of compression and mixing take us to $E_i^{is}(T, p, x_i)$. The non-ideality property is the last increment to arrive at $E_i(T, p, x_i)$.

In this section we deal with the standard properties of heating. As an additional approximative experimental law we assume $C_{p,i}^0$ to be a linear function of T. The standard properties of heating for

$$H_i, G_i/T, \varepsilon_i \text{ and } S_i$$

are derived by applying the differentials w.r.t. T as given by the set of single-phase PR's in Sect. 3.2.3. The results in terms of the constants A_i and B_i can be rewritten as PR's which contain values of $C_{p,i}^0$ or H_i^0 at a characteristic average of the initial temperature T_0 and the final temperature T.

$$\boxed{C^0_{p,i} = A_i + B_i T}$$

$$dH^0_i = C^0_{p,i}\,dT = (A_i + B_i T)\,dT$$

$$H^0_i = H^0_i(T_0) + A_i(T - T_0) + \frac{B_i}{2}(T^2 - T_0^2)$$

$$\boxed{H^0_i = H^0_i(T_0) + C^0_{p,i}(T_1)(T - T_0)}\ ,\qquad \boxed{T_1 = \frac{T_0 + T}{2}}$$

$$d\left(\frac{G^0_i}{T}\right) = -\frac{H^0_i}{T^2}\,dT = d\left(\frac{H^0_i}{T}\right) - \frac{1}{T}\,dH^0_i = d\left(\frac{H^0_i}{T}\right) - \frac{A_i + B_i T}{T}\,dT$$

$$\frac{G^0_i}{T} = \frac{G^0_i(T_0)}{T_0} + \left\{\frac{H^0_i}{T} - \frac{H^0_i(T_0)}{T_0}\right\} - \left\{A_i \ln\frac{T}{T_0} + B_i(T - T_0)\right\}$$

$$= \frac{G^0_i(T_0)}{T_0} + H^0_i(T_0)\left(\frac{1}{T} - \frac{1}{T_0}\right) + A_i\left(\frac{T - T_0}{T} - \ln\frac{T}{T_0}\right) + B_i\left\{\frac{T^2 - T_0^2}{2T} - (T - T_0)\right\}$$

$$= \frac{G^0_i(T_0)}{T_0} + \left\{H^0_i(T_0) + A_i(T_2 - T_0) + \frac{B_i}{2}(T_0 T - T_0^2)\right\}\left(\frac{1}{T} - \frac{1}{T_0}\right)$$

$$\frac{G^0_i}{T} = \frac{G^0_i(T_0)}{T_0} + \left\{H^0_i(T_2) + \frac{B_i}{2}(T_0 T - T_2^2)\right\}\left(\frac{1}{T} - \frac{1}{T_0}\right)$$

$$\boxed{\frac{G^0_i}{T} \approx \frac{G^0_i(T_0)}{T_0} + H^0_i(T_2)\left(\frac{1}{T} - \frac{1}{T_0}\right)}\ ,\qquad \boxed{T_2 = \frac{T_0 T}{T - T_0}\ln\frac{T}{T_0}}$$

$$d\varepsilon^0_i = C^0_{p,i}\left(1 - \frac{T^*}{T}\right)dT = (A_i + B_i T)\left(1 - \frac{T^*}{T}\right)dT$$

$$\varepsilon^0_i = \varepsilon_i(T_0) + A_i\left\{(T - T_0) - T^* \ln\frac{T}{T_0}\right\} + B_i\left\{\frac{T^2 - T_0^2}{2} - T^*(T - T_0)\right\}$$

$$\boxed{\varepsilon^0_i = \varepsilon^0_i(T_0) + C^0_{p,i}(T_3)\left\{(T - T_0) - T^* \ln\frac{T}{T_0}\right\}}\ ,\qquad \boxed{T_3 = \frac{\dfrac{T_0 + T}{2} - T^*}{1 - \dfrac{T^* \ln T/T_0}{T - T_0}}}$$

$$dS^0_i = \frac{C^0_{p,i}}{T}\,dT = \frac{A_i + B_i T}{T}\,dT$$

$$S^0_i = S^0_i(T_0) + A_i \ln\frac{T}{T_0} + B_i(T - T_0)$$

$$\boxed{S^0_i = S^0_i(T_0) + C^0_{p,i}(T_4)\ln\frac{T}{T_0}}\qquad \boxed{T_4 = \frac{T - T_0}{\ln\dfrac{T}{T_0}}}$$

3.3.2 μ_i-Models

In Sect. 1.3.1 we postulated the infinite-dilution law and the ideal-gas law for the chemical potential μ_i. These phenomenological laws on asymptotic phase behaviour can be extended with the zero-dilution law. This additional law follows from the infinite-dilution law and the Gibbs–Duhem relation for G.

On the basis of the asymptotic laws three fictitious ideal states with extended asymptotic phase behaviour can be defined: the infinite-dilution, the zero-dilution and the ideal-gas states. The real-state μ_i's depart from the ideal-state μ_i's by the partial molar non-ideality free enthalpies: the excess properties $G_i^{e,\infty}$ and $G_i^{e,z}$, and the residual property G_i^r. By definition all non-ideality properties vanish at the conditions where the laws on asymptotic phase behaviour apply. The non-ideality coefficients, the activity coefficients γ_i^∞ and γ_i^z, and the fugacity coefficient φ_i become then unity.

The three ideal states have the same $\Delta G_{M,i}$. The corresponding volume change of mixing $\Delta V_{M,i}$ is zero or, in other words, the ideal-state values of V_i are independent of x_i and y_i. For the liquid-phase ideal states we shall assume that V_i is to a first approximation independent of T and p as well. The ideal-gas V_i does depend on T and p.

We are now ready to write the ideal-state μ_i's as a sum of the standard-state property, the ideal-state properties of compression and mixing, and the non-ideality property. For the standard-state pressure we choose $p_0 = 0$ for a solid or a liquid and $p_0 = 1$ for a gas.

Asymptotic laws

infinite-dilution law: $$\boxed{(d\mu_i)_{T,p} = RT\,d(\ln x_i) \quad \text{if } x_i \to 0}$$

$x_j \to 1, \quad x_{i \neq j} \to 0$ $$\sum_i x_i (d\mu_i)_{T,p} = 0, \quad \sum_{i \neq j} x_i \frac{RT}{x_i} dx_i + x_j (d\mu_j)_{T,p} = 0$$

zero-dilution law: $$\boxed{(d\mu_j)_{T,p} = RT\,d(\ln x_j) \quad \text{if } x_j \to 1}$$

ideal-gas law: $$\boxed{(d\mu_i)_T = RT\,d\ln(y_i p) \quad \text{if } p \to 0}$$

Ideal-state μ_i and non-ideality properties

μ_i^{is}	G_i^{ni}	$G_i^{ni} = 0$
$\mu_i^\infty = \bar{G}_i^\infty + RT \ln x_i$	$G_i^{e,\infty} = RT \ln \gamma_i^\infty$	$x_i \to 0$
$\mu_i^z = \bar{G}_i + RT \ln x_i$	$G_i^{e,z} = RT \ln \gamma_i^z$	$x_i \to 1$
$\mu_i^{ig} = \bar{G}_i^{ig} + RT \ln y_i$		
$= \bar{G}_i^{ig}(T, 1) + RT \ln y_i p$	$G_i^r = RT \ln \varphi_i$	$p \to 0$

all ideal states: $$\Delta G_{M,i} = RT \ln x_i, \quad \Delta V_{M,i} = \frac{\partial}{\partial p} \Delta G_{M,i} = 0$$

ideal-gas state: $$V_i = \frac{\partial \mu_i}{\partial p} = \frac{RT}{p}$$

Real-state μ_i

solute in solution: $$\mu_i(T, p, x_i) = \bar{G}_i^\infty(T, 0) + V_i^\infty p + RT \ln x_i + G_i^{e,\infty}$$

component of liquid: $$\mu_i(T, p, x_i) = \bar{G}_i(T, 0) + \bar{V}_i p + RT \ln x_i + G_i^{e,z}$$

component of gas: $$\mu_i(T, p, y_i) = \bar{G}_i^{ig}(T, 1) + RT \ln p + RT \ln y_i + G_i^r$$

3.3.3 Properties of Compression and Mixing

From the μ_i-models developed in the previous section, other useful functions can be derived merely using the set of single-phase PR's.

For H_i the ideal-state property of mixing is zero for a gas and for a liquid. The contribution of the ideal-state entropy of mixing to S_i for a gas and for a liquid is positive in line with the concept of entropy as a measure of disorder of matter. The expressions for ε_i resemble those for μ_i. Changes of μ_i/RT due to heating, compression and mixing will turn out to be particularly useful in deriving PR's of a multi-phase system.

It should be noted that we neglected the thermal expansion and compressibility of solids and ideal-state liquids in deriving the PR's presented. In most applications this is well justified.

μ_i	solid	$\bar{G}_i = G_i^0 + V_i^0 p$
	liquid	$\mu_i = G_i^0 + V_i^0 p + RT \ln x_i + G_i^e$
	gas	$\mu_i = G_i^0 + RT \ln y_i p + G_i^r$
$H_i = -T^2 \dfrac{\partial}{\partial T}\left(\dfrac{\mu_i}{T}\right)$	solid	$\bar{H}_i = H_i^0 + V_i^0 p$
	liquid	$H_i = H_i^0 + V_i^0 p + H_i^e$
	gas	$H_i = H_i^0 + H_i^r$
$S_i = -\dfrac{\partial \mu_i}{\partial T}$	solid	$\bar{S}_i = S_i^0$
	liquid	$S_i = S_i^0 - R \ln x_i + S_i^e$
	gas	$S_i = S_i^0 - R \ln y_i p + S_i^r$
$\varepsilon_i = H_i - T^* S_i$	solid	$\bar{\varepsilon}_i = \varepsilon_i^0$
	liquid	$\varepsilon_i = \varepsilon_i^0 + V_i^0 p + RT^* \ln x_i + \varepsilon_i^e$
	gas	$\varepsilon_i = \varepsilon_i^0 + RT^* \ln y_i p + \varepsilon_i^r$
$\dfrac{\mu_i}{RT}$		$d\left(\dfrac{\mu_i}{RT}\right) = -\dfrac{H_i}{RT^2} dT + \dfrac{V_i}{RT} dp + \left(d \dfrac{\mu_i}{RT}\right)_{T,p}$
	solid	$\Delta\left(\dfrac{\bar{G}_i}{RT}\right) = \dfrac{\bar{H}_i}{R} \Delta\left(\dfrac{1}{T}\right) + \dfrac{\bar{V}_i \Delta p}{RT}$
	liquid	$\Delta\left(\dfrac{\mu_i}{RT}\right) = \dfrac{H_i}{R} \Delta\left(\dfrac{1}{T}\right) + \dfrac{V_i \Delta p}{RT} + \Delta \ln(\gamma_i x_i)$
	gas	$\Delta\left(\dfrac{\mu_i}{RT}\right) = \dfrac{H_i}{R} \Delta\left(\dfrac{1}{T}\right) + \Delta \ln(\varphi_i y_i p)$

3.3.4 Exergy

How do we arrive at the partial molar exergy of a compound as a function of the phase it is in, T, p and composition at given infinite-reservoir temperature and pressure, T^* and p^*, and chosen set of compounds j in abundant supply with zero partial molar exergies?

The choice of the set of compounds j depends on the system considered. For a system in which compounds of C, H, O and N are chemically converted in each other, this set could comprise CO_2, O_2, N_2 and H_2O as occurring in the surrounding air and water.

First we derive the standard exergies $\varepsilon_l^0(T^*)$ of the elements to match the conditions of the infinite reservoir. Standard-state heating to T^* yields the standard free enthalpies of formation $\Delta G_{f,j}^0(T^*)$ from tabulated standard properties at T_0 using the PR's presented in Sect. 3.3.1
Compression and mixing of the components j in abundant supply from standard state to zero exergy give PR's from which the values of $\varepsilon_j^0(T^*)$ can be calculated. Usually the non-ideality contributions are here negligible.
Finally, standard-state formation of the compounds j at T^*, where exergy changes coincide with changes in free enthalpy, gives PR's from which the standard exergies $\varepsilon_l^0(T^*)$ of the elements can be solved. In the example above these are values for C, H_2, O_2 and N_2.

When knowing the values of $\varepsilon_l^0(T^*)$, the partial exergy of a compound i as a function of T, p and composition, $\bar{\varepsilon}_i(T, p)$, $\varepsilon_i(T, p, x_i)$ or $\varepsilon_i(T, p, y_i)$, is found by a number of consecutive steps: standard-state heating to find $\Delta G_{f,i}^0(T^*)$ from standard properties at T_0, standard formation to arrive at $\varepsilon_i^0(T^*)$, standard heating from T^* to T to come to $\varepsilon_i^0(T)$, and, finally, compression and mixing.

Components in abundant supply

standard-state heating to T*:

$$\frac{\Delta G^0_{f,j}(T^*)}{T^*} = \frac{\Delta G^0_{f,j}(T_0)}{T_0} + \Big\{\Delta H^0_{f,j}(T_0) + \Delta A_{f,j}(T_2 - T_0) + \frac{\Delta B_{f,j}}{2}(T^* T_0 - T_0^2)\Big\}\left(\frac{1}{T^*} - \frac{1}{T_0}\right)$$

compression and mixing to zero-exergy state:

solid: $0 = \varepsilon_j^0(T^*) + V_j^0 p^*$

liquid: $0 = \varepsilon_j^0(T^*) + V_j^0 p^* + RT^* \ln \gamma_j^* x_j^*$

gas: $0 = \varepsilon_j^0(T^*) + RT^* \ln \varphi_j^* y_j^* p^*$

standard–state formation:

$$\varepsilon_j^0(T^*) = \sum b_{lj} \varepsilon_l^0(T^*) + \Delta G^0_{f,j}(T^*)$$

Partial molar exergy $\varepsilon_i(T, p, x_i)$

$$\frac{\Delta G^0_{f,i}(T^*)}{T^*} = \frac{\Delta G^0_{f,i}(T_0)}{T_0} + \Big\{\Delta H_{f,i}(T_0) + \Delta A_{f,i}(T_2 - T_0) + \frac{\Delta B_{f,i}}{2}(T^* T_0 - T_0^2)\Big\}\left(\frac{1}{T^*} - \frac{1}{T_0}\right)$$

$$\varepsilon_i^0(T^*) = \sum b_{li} \varepsilon_l^0(T^*) + \Delta G^0_{f,i}(T^*)$$

$$\varepsilon_i^0(T) = \varepsilon_i^0(T^*) + A_i\Big\{(T - T^*) - T^* \ln \frac{T}{T^*}\Big\} + \frac{B_i}{2}(T - T^*)^2$$

solid: $\bar{\varepsilon}_i(T, p) = \varepsilon_i^0(T) + V_i^0 p$

liquid: $\varepsilon_i(T, p, x_i) = \varepsilon_i^0(T) + V_i^0 p + RT^* \ln x_i + (H_i^e - T^* S_i^e)$

gas: $\varepsilon_i(T, p, y_i) = \varepsilon_i^0(T) + RT^* \ln y_i p + (H_i^r - T^* S_i^r)$

3.3.5 Closed Ideal-Gas System

In this section we shall develop ideal-gas PR's including some already encountered.

First we insert the ideal-gas T-p-$\bar{V}$ behaviour into the functions of measurables and PR's between measurables as derived in Sect. 3.2.2. $\bar{H}$, $\bar{C}_p$, $\bar{U}$ ad $\bar{C}_v$ appear to be functions of T only. The Joule–Kelvin and the Joule effects are zero. $\bar{C}_p$ and $\bar{C}_v$ of an ideal gas are interconnected by a simple PR as are K_T and K_S, the isentropic compressibility being the smaller of the two. PR's with one intensive kept constant follow immediately as special cases. The ratio of the heat capacities, κ occurring in the isentropic T-p, T-$\bar{V}$ and p-$\bar{V}$ relations is assumed to be independent of T.

Functions of measurables and PR's between measurables

$\bar{V} = \frac{RT}{p}$ $d\bar{S} = \frac{\bar{C}_p}{T} dT - \frac{R}{p} dp$ $d\bar{H} = \bar{C}_p dT$	$\alpha = \frac{1}{\bar{V}} \frac{\partial \bar{V}}{\partial T} = \frac{1}{T}, \quad K_T = -\frac{1}{\bar{V}} \frac{\partial \bar{V}}{\partial p} = \frac{1}{p}$ $\frac{\partial \bar{C}_p}{\partial p} = 0$ $\left(\frac{\partial T}{\partial p}\right)_{\bar{H}} = 0$ (Joule–Kelvin)
$p = \frac{RT}{\bar{V}}$ $d\bar{S} = \frac{\bar{C}_v}{T} dT + \frac{R}{\bar{V}} d\bar{V}$ $d\bar{U} = \bar{C}_v dT$	$\frac{\partial \bar{C}_v}{\partial \bar{V}} = 0 \qquad \bar{C}_p - \bar{C}_v = R$ $\left(\frac{\partial T}{\partial \bar{V}}\right)_{\bar{U}} = 0$ (Joule)
$T = \frac{p\bar{V}}{R}$ $d\bar{S} = \frac{\bar{C}_v}{p} dp + \frac{\bar{C}_p}{\bar{V}} d\bar{V}$	$K_s = \frac{\bar{C}_v}{\bar{C}_p} K_T$

PR's with one intensive kept constant $\left(\kappa = \frac{\bar{C}_p}{\bar{C}_v}\right)$

isotherm, $dT = 0$	isentrope, $d\bar{S} = 0$	isochor, $d\bar{V} = 0$	isobar, $dp = 0$
	$Tp^{-\frac{\kappa - 1}{\kappa}} = \text{constant}$	$Tp^{-1} = \text{constant}$	
	$T\bar{V}^{\kappa - 1} = \text{constant}$		$T\bar{V}^{-1} = \text{constant}$
$p\bar{V} = \text{constant}$	$p\bar{V}^{\kappa} = \text{constant}$		
$d\bar{S} = -R d\ln p$ $= R d\ln \bar{V}$		$d\bar{S} = \bar{C}_v d\ln T$	$d\bar{S} = \bar{C}_p d\ln T$
$d\bar{U} = d\bar{H} = 0$	$d\bar{U} = \bar{C}_v dT$ $d\bar{H} = \bar{C}_p dT$	$d\bar{U} = \bar{C}_v dT$	$d\bar{H} = \bar{C}_p dT$
$d\bar{F} = d\bar{G} = RT d\ln p$ $= -RT d\ln \bar{V}$			

Another route in developing ideal-gas PR's is to start from the expression for $\bar{G}$. The properties of compression for $\bar{U}$ and $\bar{H}$ are again found to be zero.

Properties of compression

$$\bar{G} = \bar{G}(p = 1) + RT\ln p = \bar{U} + p\bar{V} - T\bar{S}$$

$$\bar{V} = \frac{\partial \bar{G}}{\partial p} = \frac{RT}{p}$$

$$\bar{S} = -\frac{\partial \bar{G}}{\partial T} = \bar{S}(1) - R\ln p$$

$$\bar{U} = \bar{G} - p\bar{V} + T\bar{S} = \bar{U}(1)$$

$$\bar{H} = \bar{G} + T\bar{S} = \bar{H}(1)$$

$$\bar{F} = \bar{U} - T\bar{S} = \bar{F}(1) + RT\ln p$$

3.3.6 Residual Properties

The residual properties of a gas, the increments to the ideal-gas properties to obtain the real-state properties, all vanish when p approaches zero or when V goes to infinity. The residual functions of state can again be expressed as functions of measurables.

From volume-explicit T-p-$\bar{V}$ and T-p-V_i data residual properties are obtained by applying equations from the set of single-phase PR's.

Alternatively, knowledge of the average molar compressibility factor $\bar{Z}(T, \bar{V})$ can be used to find the average molar residual properties. From $Z(T, V, n_i)$ the residual chemical potential μ_i^r and, hence, the fugacity coefficient φ_i can be computed.

Functions of T and p

$p\bar{V}^r = p\bar{V} - RT$	$pV_i^r = pV_i - RT$
$\bar{G}^r = RT\ln\bar{\varphi} = \int_0^p \bar{V}^r\,dp$ $\bar{S}^r = -\frac{\partial\bar{G}^r}{\partial T} = -\int_0^p \frac{\partial\bar{V}^r}{\partial T}\,dp$	$G_i^r = RT\ln\varphi_i = \int_0^p V_i^r\,dp$ $S_i^r = -\frac{\partial G_i^r}{\partial T} = -\int_0^p \frac{\partial V_i^r}{\partial T}\,dp$

Functions of T and $\bar{V}$	**Functions of T, V and n_i**
$p = \frac{RT}{\bar{V}}\bar{Z}, \quad \bar{Z}(T,\infty) = 1$	$p = \frac{RT}{V}Z, \quad Z(T,\infty,n_i) = n$
$p\bar{V}^r = RT(\bar{Z}-1)$	
$\frac{\partial\bar{F}^r}{\partial\bar{V}} = \frac{\partial\bar{F}}{\partial\bar{V}} - \frac{\partial\bar{F}^{ig}}{\partial\bar{V}} =$ $= -p - RT\frac{\partial}{\partial\bar{V}}(\ln p) =$ $= -\frac{RT}{\bar{V}}(\bar{Z}-1) - RT\frac{\partial}{\partial\bar{V}}(\ln\bar{Z})$	$\frac{\partial\mu_i^r}{\partial V} = \frac{\partial\mu_i}{\partial V} - \frac{\partial\mu_i^{ig}}{\partial V} =$ $= -\frac{\partial p}{\partial n_i} - RT\frac{\partial}{\partial V}(\ln p)$ $= -\frac{RT}{V}\left(\frac{\partial Z}{\partial n_i} - 1\right) - RT\frac{\partial}{\partial V}(\ln Z)$
$\bar{F}^r = -\int_\infty^{\bar{V}} RT(\bar{Z}-1)\frac{d\bar{V}}{\bar{V}} - RT\ln\bar{Z}$	$\mu_i^r = -\int_\infty^{V} RT\left(\frac{\partial Z}{\partial n_i} - 1\right)\frac{dV}{V} - RT\ln\bar{Z}$
$\frac{\partial\bar{U}^r}{\partial\bar{V}} = \frac{\partial\bar{U}}{\partial\bar{V}} - \frac{\partial\bar{U}^{ig}}{\partial\bar{V}} = T\frac{\partial p}{\partial T} - p$ $= \frac{RT^2}{\bar{V}}\frac{\partial\bar{Z}}{\partial T}$	
$\bar{U}^r = RT^2\int_\infty^{\bar{V}} \frac{\partial\bar{Z}}{\partial T}\frac{d\bar{V}}{\bar{V}}$	

3.3.7 Activity Coefficients

The activity coefficients of the components of a liquid are interrelated by the Gibbs–Duhem relation for G. Here the relation is given for a solvent and its solutes. For an electrolyte solution and the activity coefficients of the dissociated electrically charged species occur as a fixed combination in the Gibbs–Duhem relation. They can therefore be combined to a single activity coefficient $\gamma_{\pm}^{\infty}$.

solvent: $\mu_j = \bar{G}_j + RT\ln\gamma_j^z x_j$

solutes: $\mu_i = \bar{G}_i^\infty + RT\ln\gamma_i^\infty x_i$

Gibbs–Duhem relation: $x_j\,d\mu_j + \sum_{\neq j} x_i\,d\mu_i = 0$

$$\boxed{x_j\,d\ln\gamma_j^z + \sum_{\neq j} x_i\,d\ln\gamma_i^\infty = 0}$$

electrolyte solution: $1\text{ mole } K_{\nu^+}A_{\nu^-} \rightarrow \nu_+ \text{ mole } K^{+z+} + \nu_- \text{ mole } A^{-z-}$

$$x_+ = \nu_+ x_\pm, \quad x_- = \nu_- x_\pm, \quad x_+ + x_- = \nu x_\pm$$

$$\boxed{\nu_+ + \nu_- \equiv \nu}$$

Gibbs–Duhem relation: $x_j\,d\ln\gamma_j^z + x_i\,d\ln\gamma_i^\infty + \nu_+ x_\pm\,d\ln\gamma_+^\infty + \nu_- x_\pm\,d\ln\gamma_-^\infty = 0$

$$\boxed{\nu_+\ln\gamma_+^\infty + \nu_-\ln\gamma_-^\infty \equiv \nu\ln\gamma_\pm^\infty}$$

$$x_j\,d\ln\gamma_j^z + x_i\,d\ln\gamma_i^\infty + \nu x_\pm\,d\ln\gamma_\pm^\infty = 0$$

strong electrolyte: $\boxed{x_j\,d\ln\gamma_j^z + \nu x_\pm\,d\ln\gamma_\pm^\infty = 0}$

As alternative composition parameters for the solutes the concentrations c_i or the molalities l_i can be used instead of the mole fractions x_i. New infinite-dilution values of μ_i at reference composition and new activity coefficients occur in the expressions for μ_i. They can easily be expressed in terms of the corresponding values based on x_i as composition parameter.
The Gibbs–Duhem relation between the activity coefficients has to be adjusted as well when using concentrations or molalities as composition parameters.

Concentration basis	**Molality basis**
$c_i = x_i c$	solutes: $l_i = \frac{n_i}{m_j} = \frac{x_i}{M_j x_j}$
solutes:	
$\boxed{\mu_i = \mu_i^\infty(c_i = 1) + RT\ln\gamma_i^{\infty,c} c_i}$	$\boxed{\mu_i = \mu_i^\infty(l_i = 1) + RT\ln\gamma_i^{\infty,l} l_i}$
$\mu_i = \bar{G}_i^\infty + RT\ln\gamma_i^\infty x_i$	$\mu_i = \bar{G}_i^\infty + RT\ln\gamma_i^\infty x_i$
$\mu_i^\infty(c_i=1) - \bar{G}_i^\infty + RT\ln\frac{\gamma_i^{\infty,c} c}{\gamma_i^\infty} \equiv 0$	$\mu_i^\infty(l_i = 1) - \bar{G}_i^\infty + RT\ln\frac{\gamma_i^{\infty,1}}{\gamma_i^\infty M_j x_j} \equiv 0$
$\boxed{\mu_i^\infty(c_i = 1) = \bar{G}_i^\infty - RT\ln\bar{c}_j}$	$\boxed{\mu_i^\infty(l_i = 1) = \bar{G}_i^\infty + RT\ln M_j}$
$\boxed{\gamma_i^{\infty,c} = \frac{\bar{c}_j}{c}\gamma_i^\infty}$	$\boxed{\gamma_i^{\infty,1} = x_j\gamma_i^\infty}$
Gibbs–Duhem relation:	Gibbs–Duhem relation:
$x_j d\mu_j + \sum_{\neq j} x_i d\mu_i = 0$	
$c_j d\ln(\gamma_j^z x_j) + \sum_{\neq j} c_i d\ln(\gamma_i^{\infty,c} c_i) = 0$	$d\ln(\gamma_j^z x_j) + \sum_{\neq j} M_j l_i d\ln(\gamma_i^{\infty,1} l_i) = 0$
strong electrolyte:	strong electrolyte:
$\boxed{c_j d\ln(\gamma_j^z x_j) + \nu c_\pm d\ln(\gamma_\pm^{\infty,c} c_\pm) = 0}$	$\boxed{d\ln(\gamma_j^z x_j) + M_j \nu l_\pm d\ln(\gamma_\pm^{\infty,1} l_\pm) = 0}$

4 PR's of Multi-Phase Systems

The development of PR's in this chapter follows the same pattern as in Chapter 3. Again inputs to each cell are from cells with equal or lower row and column numbers.

4.1 Column-1 PR's of Multi-Phase Systems

For a system of ω coexistent phases of given intensive states ω extensives are needed to specify the system completely. When the system is closed the N balance equations $dn_i = 0$ reduce this number to $f_{ext} = \omega - N$. When we consider a closed system with $f_{int} = f_{ext} = 1$ and select T as the intensive independent and n^β as the extensive independent property, the generalised property differential $dE(T, n^\beta)$ can be formulated. It contains the phase-transition property $\Delta\bar{E}$. For $\omega = 3$ and $N = 2$ the lever rule gives the mole ratios at which the phases α and γ form phase β of intermediate composition or vice versa.

4.2 Column-2 PR's of Multi-Phase Systems

Inputs to the PR's developed in this cell are in addition to the total property differential $dE(T, n^\beta)$ for a closed system as formulated in cell 4.1, the interface equilibrium conditions, the FPR of matter and the FBE's for reversible heating as derived in Sect. 2.2.5 along with some column-2 single-phase PR's.

First we establish that the number of intensive independents of a multi-phase system is given by Gibbs phase rule: $f_{int} = 2 + N - \omega$.

For a closed multi-phase system we find $f = 2$ for the number of independent properties and $f_{ext} = \omega - N$ for the number of extensive independents.

For a closed system with $f_{int} = f_{ext} = 1$ reversible heating connects dQ with the properties of phase transition $\Delta\bar{H}$ and $\Delta\bar{S}$. The free enthalpy of phase transition $\Delta\bar{G} = 0$. The Clausius–Clapeyron equation

$$\frac{dp}{dT} = \frac{\Delta\bar{S}}{\Delta\bar{V}} = \frac{\Delta\bar{H}}{T\Delta\bar{V}}$$

connects measurable properties.

For $\omega = 2$ and $N = 1$, the Planck equation expresses the change of $\Delta\bar{H}$ with T in terms of other phase-transition properties.

For a two-phase binary system ($\omega = 2$ and $N = 2$) $f_{int} = 2$. Consequently a PR must exist between for example T, p and x^{α}, the mole fraction of one of the components in one of the phases. This PR reduces to the Clausius–Clapeyron equation for an azeotrope at equal compositions in vapour and liquid phases.

4.3 Column-3 PR's of Multi-Phase Systems

Here we use in addition to the findings in cell 4.2, the interface equilibrium conditions derived in cell 2.2 and the property increments presented in cell 3.3.

First we classify single-component systems according to the number of coexistent phases with the aid of the Gibbs phase rule. The systems with $\omega = 1$, 2 or 3 are represented as projections of p-T-$\bar{V}$ space.

Binary systems are classified along the same lines. The various systems with $\omega = 1$, 2, 3 or 4 are represented as projections of p-T-x space.

We consider G-S, L-S, G-L and L-L systems. The PR's developed for these two-phase systems involve measurable properties, single-phase non-ideality coefficients and properties of phase transition. They include expressions for sublimation-point depression, melting-point depression, solubility of a solid, boiling point elevation by a less volatile component, and osmotic pressure of a solution as well as the laws of Raoult, Henry and Nernst.

4.1.1 Closed System with $f_{int} = f_{ext} = 1$

For ω coexistent phases of given intensive states ω extensives are needed to completely specify the system. When the system is closed, N balance equations $dn_i = 0$ reduce this number to $f_{ext} = \omega - N$.

Here we restrict ourselves to a system with $f_{ext} = 1$ or $\omega - N = 1$. As a closed system has in total two degrees of freedom (see next section), the system must have one intensive degree of freedom: $f_{int} = 1$. Accordingly we may select T and n^β, the mole number of one of the phases, as the independents of the system and formulate the generalised property differential $dE(T, n^\beta)$ containing the phase-transition property $\Delta\bar{E}$.

To the $\omega - 1$ dependent n^α's phase-transition coefficients ν^α can be attached in addition to $\nu^\beta \equiv 1$. They can be calculated from the mole balances. For a system with $\omega = 2$ and $N = 1$ we find $\nu^\alpha = -1$ and $\nu^\beta = +1$. For $\omega = 3$ and $N = 2$ the mole ratios at which the phases α and γ are converted into phase β of intermediate composition, or vice versa, are given by the lever rule. The property of phase transition $\Delta\bar{E}$ can readily be expressed in terms of the ν^α's.

ω n^α's as primary extensives
N BE's $\quad dn_i = 0$

$$\boxed{f_{ext} = \omega - N}$$

<u>$f_{int} = f_{ext} = 1$:</u>

$$\boxed{dE = \frac{\partial E}{\partial T} dT + \Delta\bar{E}\, dn^\beta}$$

n^α	$\frac{\partial n^\alpha}{\partial n^\beta} \equiv \nu^\alpha$	$\nu^\beta \equiv 1$	
$n = \sum n^\alpha = \sum n_i$	$\frac{\partial n}{\partial n^\beta} = 0$	$\sum \nu^\alpha = 0$	$\boxed{\sum_{\neq\beta} (-\nu^\alpha) = 1}$
$n_i = \sum x_i^\alpha n^\alpha$	$\frac{\partial n_i}{\partial n^\beta} = 0$	$\sum x_i^\alpha \nu^\alpha = \sum (x_i^\alpha - x_i^\beta)\nu^\alpha = 0$	$\boxed{\sum_{\neq\beta} (x_i^\beta - x_i^\alpha)(-\nu^\alpha) = 0}$
		$\omega = 2, \quad N = 1$: $\nu^\beta = 1, \ -\nu^\alpha = 1$	

$\omega = 3, \quad N = 2$:

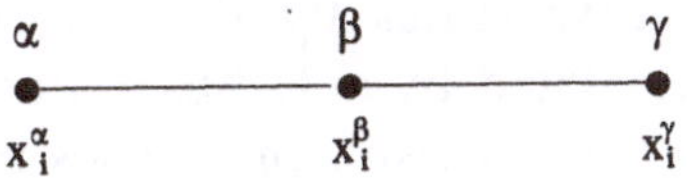

$\nu^\beta = 1, \quad (-\nu^\alpha) + (-\nu^\gamma) = 1$

lever rule: $\boxed{(x_i^\beta - x_i^\alpha)(-\nu^\alpha) = (x_i^\gamma - x_i^\beta)(-\nu^\gamma)}$

$E = \sum n^\alpha \bar{E}^\alpha$	$\frac{\partial E}{\partial n^\beta} = \Delta\bar{E}$	$\Delta\bar{E} = \sum \nu^\alpha \bar{E}^\alpha = \sum (\bar{E}^\alpha - \bar{E}^\beta)\nu^\alpha$	$\boxed{\Delta\bar{E} = \sum_{\neq\beta} (\bar{E}^\beta - \bar{E}^\alpha)(-\nu^\alpha)}$

4.2.1 FPR and Equivalents. Degrees of Freedom. Gibbs Phase Rule

As for a single-phase system three alternative representations of the FPR of matter can be formulated. One of the cross-differentiation identities of dF will be used in the next section.

The number of independent properties or degrees of freedom determining an open multi-phase system follows from the FPR: $f = 2 + N$. To describe the extensive states of the phases at given intensive states ω mole numbers, one for each phase, are needed: $f_{ext} = \omega$. The resultant number of intensive independents determining all coexistent intensive states is given by the Gibbs phase rule.

For a closed system $dn_i = 0$ for each of the N components. Accordingly the number of independents becomes N fewer than for an open system: $f = 2$.
If we describe the extensive states at given intensive states by the ω n^{α}'s, the N balance equations $dn_i = 0$ of a closed system reduce the number of independent n^{α}'s to $f_{ext} = \omega - N$. The resultant number of intensive independents is again given by the Gibbs phase rule.

Yet another way to derive the Gibbs phase rule is to describe the intensive states by the $2 + N$ intensives T, p and N μ_i's. These are interrelated by ω Gibbs–Duhem relations, so that the number of independent intensive properties becomes $f_{int} = 2 + N - \omega$.

U	$dU = \quad T\,dS - p\,dV + \sum \mu_i\,dn_i$	
$H = U + pV$	$dH = \quad T\,dS + V\,dp + \sum \mu_i\,dn_i$	
$F = U - TS$	$dF = -S\,dT - p\,dV + \sum \mu_i\,dn_i$	$\left(\frac{\partial p}{\partial T}\right)_{V,n_i} = \left(\frac{\partial S}{\partial V}\right)_{T,n_i}$
$G = U + pV - TS$	$dG = -S\,dT + V\,dp + \sum \mu_i\,dn_i$	

reversible heating: $dQ = dU + p\,dV = T\,dS$
(closed system)

Degrees of freedom

	extensive and intensive states of coexistent phases		intensive states
	number of independents f	minimum number of extensive ind. f_{ext}	number of intensive ind. $f_{int} = f - f_{ext}$
open system	$f = 2 + N$	$f_{ext} = \omega$	$f_{int} = 2 + N - \omega$
closed system $dn_i = 0$ or $\sum x_i^\alpha dn^\alpha = 0$	$f = 2$	$f_{ext} = \omega - N$	$f_{int} = 2 - (\omega - N)$

Gibbs phase rule

interface equilibria: $T^\alpha = T^\beta = T$, $\quad p^\alpha = p^\beta = p$, $\quad \mu_i^\alpha = \mu_i^\beta = \mu_i$

$2 + N$ primary intensives: T, p, $N\,\mu_i$'s

ω restrictive equations: $\sum x_i^\alpha\,d\mu_i = -\bar{S}^\alpha\,dT + \bar{V}^\alpha\,dp$

(Gibbs–Duhem relations)

$$\boxed{f_{int} = 2 + N - \omega}$$

4.2.2 Closed System with $f_{int} = f_{ext} = 1$

We shall apply the PR's found in the previous sections to a system with $f_{int} = f_{ext} = 1$. For such a system p and any other intensive can be expressed as a function of T only.

The free enthalpy of phase transition $\Delta\bar{G}$ appears to be zero. Isobaric or isothermal heating relates dQ to $\Delta\bar{H}$ and $\Delta\bar{S}$. From experimental experience it can be concluded that H and S increase in the order solid-liquid-gas. This is again in agreement with the concept of entropy as a measure of disorder.

The derivative of p as a function of T for a system with $f_{int} = f_{ext} = 1$ is given by the Clausius–Clapeyron equation. Derivations of this equation by three routes are indicated: from a cross-differentiation identity of $dF(T, V, n_i)$, from the summation of the Gibbs–Duhem relations and from the cross-differentiation identity of $dG(T, n^{\beta}) = dG(T, p)$.

The Planck equation of a two-phase single-component system follows from $d\bar{H}(T, p)$ for the individual phases and the Clausius–Clapeyron equation. It relates properties of phase transition. When one of the phases is a gas which approaches the ideal-gas state and when the molar volume of the condensed phase is neglected w.r.t. that of the gas phase, the Planck equation can be further simplified to a relation between $\Delta\bar{H}$ and $\Delta\bar{C}_p$.

Phase-transition properties

$$\Delta\bar{G} = \sum \nu^{\alpha}\bar{G}^{\alpha} = \sum\sum \nu^{\alpha}x_i^{\alpha}\mu_i = 0 \qquad \boxed{\Delta\bar{G} = 0, \quad \Delta\bar{H} - T\,\Delta\bar{S} = 0}$$

$$(dQ)_p = \Delta\bar{H}\,dn^{\beta} = T\Delta\bar{S}\,dn^{\beta}$$

Clausius–Clapeyron equation

from cross-differentiation identity of $dF(T, V, n_i)$:

$$\left(\frac{\partial p}{\partial T}\right)_{V,n_i} = \left(\frac{\partial S}{\partial V}\right)_{T,n_i} \qquad \boxed{\frac{dp}{dT} = \frac{\Delta\bar{S}}{\Delta\bar{V}} = \frac{\Delta\bar{H}}{T\Delta\bar{V}}}$$

from Gibbs–Duhem relations:

$$\sum x_i^{\alpha}\,d\mu_i = -\,\bar{S}^{\alpha}\,dT + \bar{V}^{\alpha}\,dp$$

$$\sum\sum \nu^{\alpha}x_i^{\alpha}\,d\mu_i = -\,\Delta\bar{S}\,dT + \Delta\bar{V}\,dp = 0$$

from cross-differentiation identity of $dG(T, n^{\beta}) = dG(T, p)$:

$$dG = -\,S\,dT + V\,dp = \left(-\,S + V\frac{dp}{dT}\right)dT$$

$$\Delta\bar{G} = \frac{\partial G}{\partial n^{\beta}} = 0 \qquad \frac{\partial^2 G}{\partial T\partial n^{\beta}} = -\,\Delta\bar{S} + \Delta\bar{V}\frac{dp}{dT} = 0$$

Planck equation: $\omega = 2$, $N = 1$

$$d(\Delta\bar{H}) = \Delta\bar{C}_p\,dT + \left(\Delta\bar{V} - T\frac{\partial}{\partial T}\Delta\bar{V}\right)dp$$

$$\boxed{\frac{d\Delta\bar{H}}{dT} = \Delta\bar{C}_p + \frac{\Delta\bar{H}}{T} - \Delta\bar{H}\frac{\partial}{\partial T}(\ln\Delta\bar{V})}$$

ideal gas: $\Delta\bar{V} \approx \dfrac{RT}{p}$, $\dfrac{d\Delta\bar{H}}{dT} \approx \Delta\bar{C}_p$

4.2.3 Two-Phase Binary System

For a two-phase binary system we find $f_{int} = 2$: any intensive can be written as a function of two intensives, say T and p.

For example, the PR between $d\mu_2$, dT and dp is found by combining the Gibbs–Duhem relations for the two phases. The composition of a binary phase is denoted by $x \equiv x_2 = 1 - x_1$. For an azeotrope, a G-L binary which can exhibit equal compositions of the coexistent phases ($x^\beta = x^\alpha$), the Clausius–Clapeyron equation emerges again.

By substituting $d\mu_2(T, p, x^\alpha)$ we find a PR between dx^α, dT and dp. For an azeotrope T as a function of composition at constant p passes through a minimum or a maximum. The same holds for p at constant T.

$$f_{int} = 2 + N - \omega =$$
$$= 2 + 2 - 2 = 2$$

Gibbs–Duhem relations

$$(1 - x^\beta)\,d\mu_1 + x^\beta\,d\mu_2 = -\bar{S}^\beta\,dT + \bar{V}^\beta\,dp$$
$$(1 - x^\alpha)\,d\mu_1 + x^\alpha\,d\mu_2 = -\bar{S}^\alpha\,dT + \bar{V}^\alpha\,dp$$

PR between $d\mu_2$, dT and dp

$$\boxed{(x^\beta - x^\alpha)\,d\mu_2 = -\Delta\bar{S}'\,dT + \Delta\bar{V}'\,dp} \qquad \Delta\bar{E}' = (1 - x^\alpha)\bar{E}^\beta - (1 - x^\beta)\bar{E}^\alpha$$

$$x^\beta = x^\alpha: \quad \frac{dp}{dT} = \frac{\Delta\bar{S}}{\Delta\bar{V}}, \qquad \Delta\bar{E} = \bar{E}^\beta - \bar{E}^\alpha$$

PR between dx^α, dT and dp

$$d\mu_2 = -S_2^\alpha\,dT + V_2^\alpha\,dp + \frac{\partial\mu_2}{\partial x^\alpha}\,dx^\alpha$$

$$\boxed{(x^\beta - x^\alpha)\frac{\partial\mu_2}{\partial x^\alpha}\,dx^\alpha = -\Delta\bar{S}''\,dT + \Delta\bar{V}''\,dp} \qquad \Delta\bar{E}'' = \Delta\bar{E}' - (x^\beta - x^\alpha)E_2^\alpha$$

$$x^\beta = x^\alpha: \quad \left(\frac{\partial T}{\partial x^\alpha}\right)_p = 0, \quad \left(\frac{\partial p}{\partial x^\alpha}\right)_T = 0$$

4.3.1 Single-Component Systems

We have completed the development of column-2 multi-phase PR's and shall as a next step embark upon the derivation of column-3 multi-phase PR's using column-3 single-phase PR's. In addition single-component and binary systems will be classified according to the number of coexistent phases ω. First we explore single-component systems comprising a combination of a gas phase G, a liquid phase L and a solid phase S.

For $\omega = 3$ we find $f_{int} = 0$. At the triple point none of the intensives of the three coexistent phases can be chosen freely. In a p-T-$\bar{V}$ space this is represented by a triplet of points with the same coordinates for T and p. For a closed system $f_{ext} = 2$: two of the three phase mole numbers n^G, n^L and n^S determine the third.

For $\omega = 2$ the number of intensive independents becomes $f_{int} = 1$. A phase in coexistence with another has one intensive degree of freedom. This results in three pairs of two-phase lines in the p-T-$\bar{V}$ space. For the G-L system the two phase lines meet at the critical point where G and L become identical. For a closed system $f_{ext} = 1$. The mole number of one of the phases fixes the other.

For $\omega = 1$, finally, we find $f_{int} = 2$. As two intensives determine a third this leads to three single-phase surfaces in the p-T-$\bar{V}$ space. For a closed system $f_{ext} = 0$. Fixing the two intensive independents specifies the extensive state as well.

From the projections of p-T-$\bar{V}$ space of a single-component system the three pairs of two-phase lines can be seen to originate from the triplet of three-phase points. Tie-lines for each pair of two-phase lines connect coexistent phases. The two-phase lines for the G-L system meet and end at the critical point. The p-T projections of each pair of two-phase lines coincide.

For the three two-phase systems with $f_{int} = 1$, PR's involving standard-state properties are found by inserting single-phase properties of compression. Yet another derivation of the Clausius–Clapeyron equation is found by combining $d(\bar{G}_i/T)$ for the coexistent phases with the equilibrium condition. As $\bar{H}_i$ increases in the order solid-liquid-gas and $\bar{V}_i$ exhibits in general the same behaviour, the slopes of the p-T curves for the three two-phase equilibria are in general positive. The L-S curve for water is one of the exceptions (ice shrinks upon melting). The L-S curve

continued on page 106

ω	f_{int}	f_{ext}	system	representation in p-T-$\bar{V}$ space
3	0	2	G-L-S (triple point) G ≡ L (critical point)	1 triplet of points 1 point
2	1	1	L-S, G-S, L-G	3 pairs of 2-phase lines
1	2	0	G, L, S	3 single-phase surfaces

$\omega = 2$

equilibrium condition: $\boxed{\Delta\bar{G}_i(T, \bar{p}_i) = 0}$ $\boxed{\Delta\bar{H}_i - T\Delta\bar{S}_i = 0}$

Clausius–Clapeyron equation

$$d\left(\frac{\Delta\bar{G}_i}{T}\right) = -\frac{\Delta\bar{H}_i}{T^2}dT + \frac{\Delta\bar{V}_i}{T}d\bar{p}_i = 0$$

$$\boxed{\frac{d\bar{p}_i}{dT} = \frac{\Delta\bar{H}_i}{T\Delta\bar{V}_i}}$$

	L-S	G-S, G-L
compression of standard states	$\boxed{\Delta G_i^0 + \Delta V_i^0 \bar{p}_i = 0}$	$\boxed{\Delta G_i^0 + RT\ln\varphi_i\bar{p}_i = V_i^0\bar{p}_i}$ $\Delta\bar{V}_i \approx \bar{V}_i^G \approx \frac{RT}{\bar{p}_i}$ $\boxed{\frac{d(\ln\bar{p}_i)}{dT} \approx \frac{\Delta\bar{H}_i}{RT^2}}$

continued on page 107

has the steepest slope because the volume change of melting/freezing is much smaller than the volume changes of solid vaporisation/sublimation and boiling/condensation. For the two-phase systems involving a vapour, the Clausius–Clapeyron equation can be further simplified when the ideal-gas state is approached.

Projections of p-T-$\bar{V}$ space

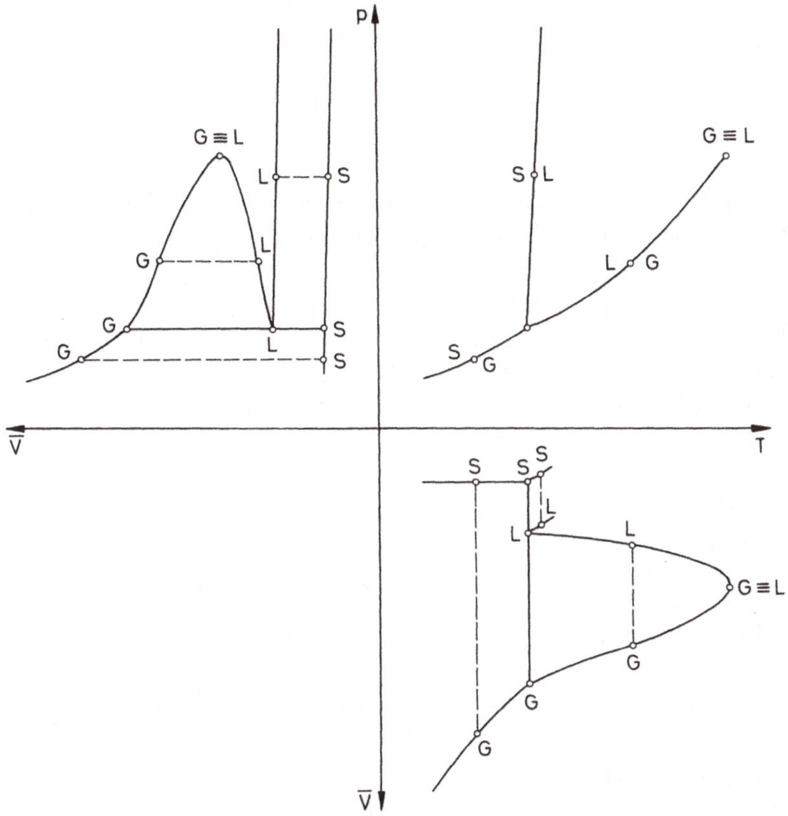

4.3.2 Binary Systems

We now focus on a binary system of the components A and B comprising a combination of S_A, L, G and S_B. The solid phases consist of pure A and pure B. We shall use $x \equiv x_B = 1 - x_A$ and y to describe the phase compositions.

For $\omega = 4$, 3, 2 and 1 we find $f_{int} = 0$, 1, 2 and 3, respectively. The resultant systems comprising combinations of S_A, G, L and S_B can be represented in a p-T-x space.
As can be seen from the p-T and T-x projections, a quadruplet of points for the four coexistent phases S_A-G-L-S_B form the origin of four triplets of lines for three coexisting phases, S_A-G-L, G-L-S_B, S_A-G-S_B and S_A-L-S_B. Coexistent phases on each triplet of three-phase lines are connected by tie-lines. The three-phase lines for S_A-G-L meet and end at the triple point of A($x = 0$), while the lines for G-L-S_B join at the triple point of B($x = 1$). Other phase lines are the critical line connecting the two-pure component critical points, and the pure-component two-phase lines at $x = 0$ and $x = 1$.
The six two-phase systems are represented by six pairs of two-phase surfaces in the p-T-x space.
Single phases, finally, give rise to regions of the p-T-x space.

For a closed system with $\omega = 4$, 3 and 2 it is found that $f_{ext} = 2$, 1 and 0, respectively.
For $\omega = 4$ two of the four phase mole numbers are independent, while the remaining two are determined by mole balances for A and B.
For a closed system with $\omega = 3$ one of the three phase mole numbers is independent, say n^{β}, β being the phase of intermediate composition. The other two phases are converted into β at fixed mole ratios given by the lever rule.
For $\omega = 2$, finally, we have no independent phase mole number left at given intensive states of the coexistent phases. At given n_A and n_B of the closed system the two phase mole numbers can be calculated with the aid of the level rule for a two-phase system.

The slopes of the p-T curves are given by the Clausius–Clapeyron equation.
The slopes for the systems which do not involve a gas phase are schematically indicated to be infinite.

continued on page 110

Classification according to ω

ω	f_{int}	f_{ext}	system	representation in p-T-x space
4	0	2	S_A-G-L-S_B (quadruple point)	quadruplet of points
3	1	1	S_A-G-L, S_A-G-S_B, S_A-L-S_B, G-L-S_B $G \equiv L$	4 triplets of 3-phase lines 1 critical line
2	2	0	S_A-G, S_A-L, S_A-S_B, G-L, G-S_B, L-S_B	6 pairs of 2-phase surfaces
1	3	—	G, L	2 single-phase regions

Slopes of p-T curves $\dfrac{dp}{dT} = \dfrac{\Delta\bar{H}}{T\Delta\bar{V}}$

$S_A \rightarrow G \leftarrow L$: $\Delta\bar{V} > 0, \quad \Delta\bar{H} > 0, \quad \dfrac{dp}{dT} > 0$

$S_A \rightarrow G \leftarrow S_B$: $\Delta\bar{V} > 0, \quad \Delta\bar{H} > 0, \quad \dfrac{dp}{dT} > 0$

$G \rightarrow L \leftarrow S_B$: $\Delta\bar{V} < 0$

$$\begin{aligned}\Delta\bar{H} &= (\bar{H}^L - \bar{H}^G)(-\nu^G) + (\bar{H}^L - \bar{H}^S)(-\nu^S)\\ &= \underset{<0 \quad \downarrow}{(\bar{H}^z - \bar{H}^G)(-\nu^G)} + \underset{>0 \quad \uparrow}{(\bar{H}^z - \bar{H}^S)(-\nu^S)} + \underset{\uparrow}{\bar{H}^e}\end{aligned}$$

$\dfrac{dp}{dT} > 0$ at quadruple point

$\dfrac{dp}{dT} < 0$ at triple point of B

continued on page 111

For S_A-G-L and S_A-G-S_B the properties of phase transition $\Delta\bar{V}$ and $\Delta\bar{H}$ have equal signs. Accordingly the slopes of the p-T curves for these systems are always positive.
For G-L-S_B, however, p as a function of T may pass through a maximum when going from the quadruple point to the triple point of B. The volume of phase transition $\Delta\bar{V}$ keeps the same sign, volume effects being dominated by the volume of the gas phase. The enthalpy of phase transition $\Delta\bar{H}$, on the other hand, may change sign due to the changes of the phase transition coefficients as given by the lever rule and the increase of the excess property $\bar{H}^e$ from a negative value to zero when the zero-dilution state is reached.

Projections of p-T-x space

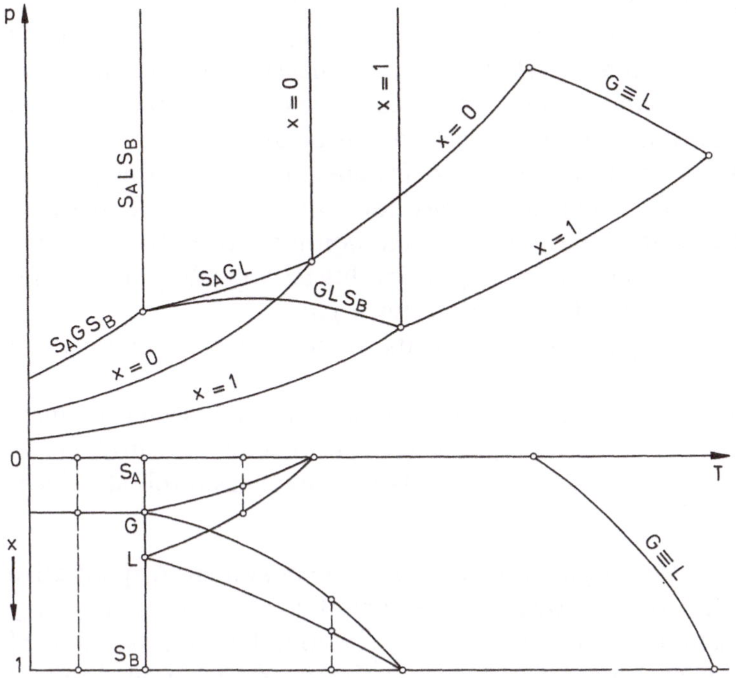

4.3.3 G-S Systems

We now start to develop PR's for the two-phase systems G-S, L-S, G-L and L-L by inserting column-3 properties of heating, compression and mixing into the column-2 equilibrium condition $\mu_i^\beta = \mu_i^\alpha$. The PR's found contain invariably measurable properties: T, p, x_i, V and associated properties, and H and associated properties measured as isobaric heat effects. Data on V and H are measured as such or extrapolated to asymptotic conditions. The PR's derived can be divided into:

- PR's which involve non-ideality coefficients and measurable properties. Under asymptotic conditions they reduce to PR's between measurables. The same occurs when non-ideality coefficients are obtained from other sources, e.g. when fugacity coefficients are calculated from measured T-p-V data for a gas.
- PR's between non-ideality coefficients, standard-state functions and measurables.

  Under asymptotic conditions they reduce to PR's which connect standard-state functions of phase transition with measurables. In turn, these can give rise to PR's involving only measurable standard-state properties.

We start with the development of PR's for a G-S system. A qualitative picture of the PR's is given by a p-y diagram at constant T and by a T-y diagram at constant p. These diagrams can be found as cross-sections of the p-T-x space at a constant T below the quadruple point and at constant p below the triple points and the quadruple point. In the cross-sections phase lines are reduced to points, phase surfaces to lines and phase regions in p-T-x space to surfaces. It can be seen that the total pressure p rises and T drops when going from pure-component G-S systems to the three-phase system S_A-G-S_B.

The PR's describing these effects are readily found by inserting properties of compression and mixing and of heating and mixing into the two-phase equilibrium condition. The asymptotic laws apply when the fugacity coefficients become unity. Usually the effect of p on $\bar{G}_i^S$ is negligible.

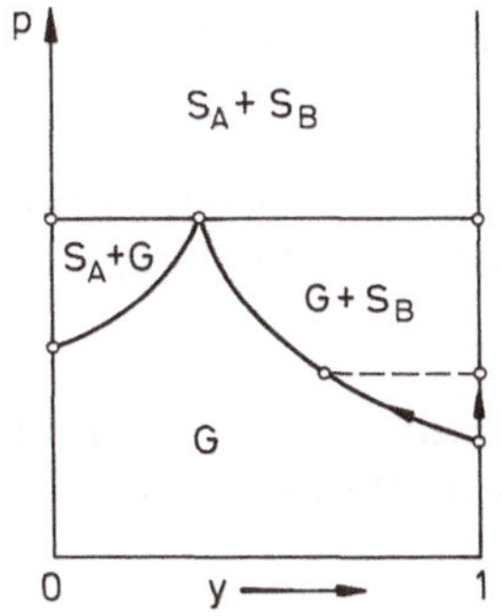

T = constant

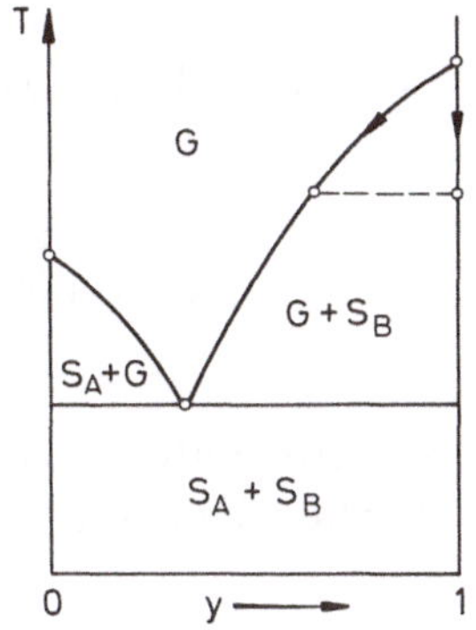

p = constant

Compression and mixing of coexistent single-component phases

$$\Delta\left(\frac{\mu_i^G}{RT}\right) = \Delta\left(\frac{\bar{G}_i^S}{RT}\right)$$

$$\ln \frac{\varphi_i y_i p}{\bar{\varphi}_i \bar{p}_i} = \frac{\bar{V}_i^S(p - \bar{p}_i)}{RT}, \quad \boxed{\frac{\varphi_i y_i p}{\bar{\varphi}_i \bar{p}_i} = \exp \frac{\bar{V}_i^S(p - \bar{p}_i)}{RT}}$$

asymptotic law: $\boxed{y_i p = \bar{p}_i}$ $\quad (1 - y)\ p = \bar{p}_A$

$$y\ p = \bar{p}_B$$

Heating and mixing of coexistent single-component phases

$$\Delta\left(\frac{\mu_i^G}{RT}\right) = \Delta\left(\frac{\bar{G}_i^S}{RT}\right)$$

$$\frac{\Delta \bar{H}_i}{R}\left(\frac{1}{T} - \frac{1}{\bar{T}_i}\right) + \ln \frac{\varphi_i y_i}{\bar{\varphi}_i} = 0$$

asymptotic law: $\boxed{\bar{T}_i - T \approx \frac{R\bar{T}_i^2}{\Delta \bar{H}_i}(1 - y_i)}$

4.3.4 L-S Systems

A cut of the p-T-x space at a constant p only intersecting the phase lines S_A-L-B, S_A-L(x = 0) and S_B-L(x = 1) illustrates the depression of a pure-component melting point by a second component.

A reduction in melting pressure at constant T can be calculated for a component like water which exhibits an anomalous volume change of melting.

Starting from standard states, a PR for the solubility of a solid in a solvent is found. The effect of p on solubility is usually negligible.

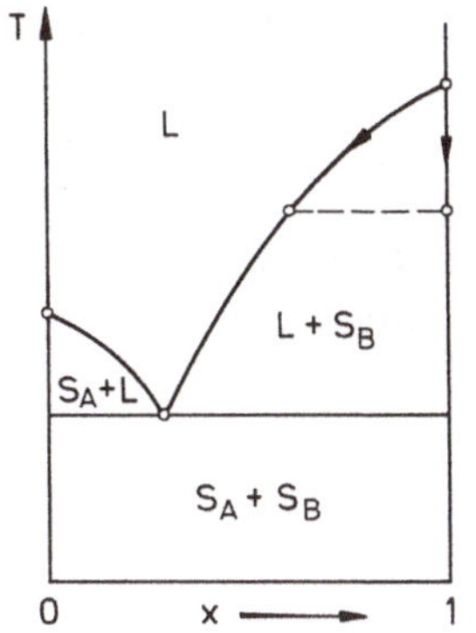

Heating and mixing of coexistent single-component phases

$$\Delta\left(\frac{\mu_i^L}{RT}\right) = \Delta\left(\frac{\bar{G}_i^S}{RT}\right)$$

$$\frac{\Delta\bar{H}_i}{R}\left(\frac{1}{T} - \frac{1}{\bar{T}_i}\right) + \ln\gamma_i^z x_i = 0 \qquad \boxed{\bar{T}_i - T \approx \frac{R\bar{T}_i^2}{\Delta\bar{H}_i}(1 - x_i)}$$

melting-point depression

Compression and mixing of coexistent single-component phases

$$\Delta\left(\frac{\mu_i^L}{RT}\right) = \Delta\left(\frac{\bar{G}_i^S}{RT}\right)$$

$$\frac{\Delta\bar{V}_i(p - \bar{p}_i)}{RT} + \ln\gamma_i^z x_i = 0 \qquad \boxed{\bar{p}_i - p = -\frac{RT}{\Delta\bar{V}_i}(1 - x_i)}$$

melting-pressure reduction
if $\bar{V}_i^L - \bar{V}_i^S < 0$

Compression and mixing of standard-state phases

$$\frac{\mu_i^L}{RT} = \frac{\bar{G}_i^S}{RT}$$

$$\frac{\Delta G_i^0}{RT} + \frac{\Delta V_i^0 p}{RT} + \ln\gamma_i^\infty x_i = 0$$

standard solubility:
$$\boxed{\ln x_i^0 = -\frac{\Delta G_i^0}{RT}}, \quad \boxed{\frac{d\ln x_i^0}{dT} = \frac{\Delta H_i^0}{RT^2}}$$

$$\boxed{\gamma_i^\infty x_i = x_i^0 \exp\left(-\frac{\Delta V_i^0 p}{RT}\right)}$$

4.3.5 G–L Systems

A cross-section of the p-T-x space at constant T which only intersects the phase lines for G–L(x = 0) and G–L(x = 1) yields a plot of p versus y and x. Two phase lines connect the pure-component boiling pressures. When comparing two coexistent vapour and liquid phases, the liquid phase contains more less-volatile component B. The G–L PR can be simplified by neglecting the effect of p on μ_i^L and reduces to Raoult's law when the non-ideality coefficients are set equal to unity.

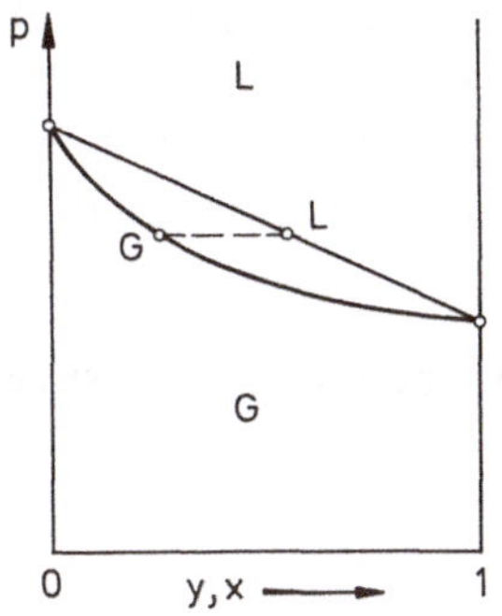

T = constant

Compression and mixing of coexistent single-component systems

$$\Delta\left(\frac{\mu_i^G}{RT}\right) = \Delta\left(\frac{\mu_i^L}{RT}\right)$$

$$\ln\frac{\varphi_i y_i p}{\bar{\varphi}_i \bar{p}_i} = \frac{\bar{V}_i^L(p - \bar{p}_i)}{RT} + \ln \gamma_i^z x_i$$

$$\boxed{\frac{y_i p}{x_i \bar{p}_i} = \frac{\gamma_i^z \bar{\varphi}_i}{\varphi_i} \exp\frac{\bar{V}_i^L(p - \bar{p}_i)}{RT}}$$

Raoult's law: $\boxed{y_i p = x_i \bar{p}_i}$

— binary system

$$(1 - y)p = (1 - x)\bar{p}_A \qquad \boxed{p = (1 - x)\bar{p}_A + x\bar{p}_B}$$

$$yp = x\bar{p}_B \qquad \boxed{\frac{1}{p} = \frac{1 - y}{\bar{p}_A} + \frac{y}{\bar{p}_B}}$$

— vapour-pressure reduction by a non-volatile

$$p = x_i \bar{p}_i = (1 - x_j)\bar{p}_i$$

A T-y, x diagram at constant p is obtained from a cross-section at constant p of the T-p-x space. The G and L two-phase lines connect the pure-component boiling points. The boiling point elevation of A by adding less-volatile component B is in the first instance proportional to x–y.

Starting from standard-state phases Henry's law for the partial pressure $y_i p$ of gas soluble in a solvent is found.

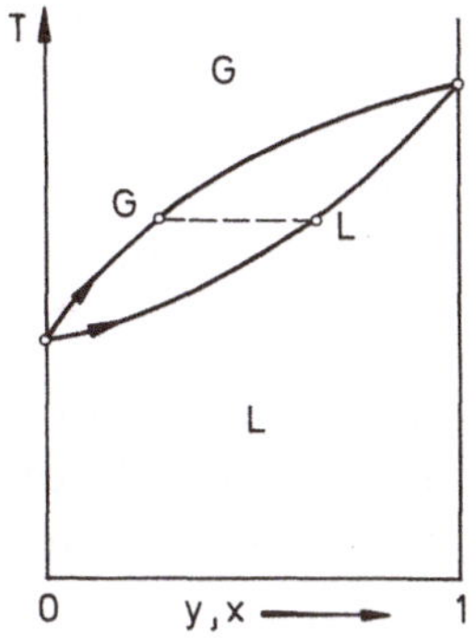

p = constant

Heating and mixing of coexistent single-component phases

$$\Delta\left(\frac{\mu_i^G}{RT}\right) = \Delta\left(\frac{\mu_i^L}{RT}\right)$$

$$\frac{\Delta\bar{\bar{H}}_i}{R}\left(\frac{1}{T} - \frac{1}{\bar{\bar{T}}_i}\right) + \ln\frac{\varphi_i y_i}{\bar{\varphi}_i} = \ln(\gamma_i^z x_i)$$

$$\frac{\Delta\bar{\bar{H}}_A}{R\bar{\bar{T}}_A^2}(\bar{T}_A - T) - y \approx -x, \qquad \boxed{T - \bar{T}_A \approx \frac{R\bar{\bar{T}}_A^2}{\Delta\bar{\bar{H}}_A}(x - y)}$$

boiling-point elevation
by less-volatile component

Compression and mixing of standard-state phases

$$\frac{\mu_i^G}{RT} = \frac{\mu_i^L}{RT}$$

$$\frac{\Delta G_i^0}{RT} + \ln\varphi_i y_i p = \frac{V_i^\infty p}{RT} + \ln\gamma_i^\infty x_i$$

standard Henry's law constant:

$$\boxed{\ln\mathscr{H}_i^0 = -\frac{\Delta G_i^0}{RT}}, \qquad \boxed{\frac{d(\ln\mathscr{H}_i^0)}{dT} = \frac{\Delta H_i^0}{RT^2}}$$

$$\boxed{\mathscr{H}_i = \frac{y_i p}{x_i} = \frac{\gamma_i^\infty \mathscr{H}_i^0}{\varphi_i}\exp\frac{V_i^\infty p}{RT}} \qquad y_i p \approx \mathscr{H}_i^0 x_i$$

4.3.6 L–L Systems

We conclude the development of PR's of two-phase systems with the derivation of PR's of L–L systems.

Nernst's law describes the distribution of a solute between two immiscible or partly miscible solvents. Again the effect of p on the μ_i's of a condensed phase is usually negligible.

A solution of i in j separated from pure j by a membrane permeable to j, has an overpressure Δp, the osmotic pressure of the solution. The PR for the osmotic pressure of a solution can be used to determine activity coefficients.

Nernst's law

compression and mixing of standard-state phases

$$\frac{\mu_i^\beta}{RT} = \frac{\mu_i^\alpha}{RT}$$

$$\frac{\Delta G_i^0}{RT} + \frac{\Delta V_i^\infty p}{RT} + \ln \gamma_i^\beta x_i^\beta = \ln \gamma_i^\alpha x_i^\alpha$$

standard Nernst's law constant

$$\boxed{\ln N_i^0 = -\frac{\Delta G_i^0}{RT}}\ , \qquad \boxed{\frac{d(\ln N_i^0)}{dT} = \frac{\Delta H_i^0}{RT^2}}$$

$$\boxed{N = \frac{x_i^\beta}{x_i^\alpha} = \frac{\gamma_i^\alpha N_i^0}{\gamma_i^\beta} \exp\left(-\frac{\Delta V_i^\infty p}{RT}\right)} \qquad x_i^\beta \approx N_i^0 x_i^\alpha$$

Osmotic pressure

$x_j = 1 - x_i$	$x_j = 1$
$p + \Delta p$	p

membrane

compression and mixing of pure j:

$$\Delta \mu_j = 0$$

$$\bar{V}_j \Delta p + RT \ln \gamma_j^z x_j = 0$$

$$\boxed{\Delta p = -\frac{RT}{\bar{V}_j} \ln \gamma_j^z (1 - x_i)} \qquad \Delta p \approx \frac{RT}{\bar{V}_j} x_i$$

5 PR's of Reaction Systems

This chapter is the last of the three chapters on PR's. The development of the PR's proceeds along the same lines as for a multi-phase system in Chapter 4.

First we formulate the column-1 total property differential $dE(T, n^\beta)$ of a closed single-reaction system prepared from a single dissociating compound. Such a system has $a = N - \omega$ additional relations between its mole fractions.

Column-2 PR's are derived using in addition to $dE(T, n^\beta)$ the findings in Sect. 2.2.6 as input.
For a closed system with $f_{int} = f_{ext} = 1$ we derive the Clausius-Clapeyron equation. Derivation and result are the same as for a non-reacting multi-phase system.
The Gibbs phase rule for the number of intensive independent properties involves the apparent number of components $N' = N - r - a$, where r denotes the number of independent reactions and a the number of additional relations e.g. the electroneutrality condition for an electrolyte solution or stoichiometric relations.

We derive column-3 PR's by substituting the μ_i-models obtained in Sect. 3.3.2 into the condition for chemical equilibrium $\sum v_i \mu_i = 0$. In the resultant column-3 equilibrium condition the standard equilibrium constant K^0 occurs. A reaction shifts further in the direction of completion, the higher K^0 and the lower the standard free enthalpy of reaction ΔG^0.
The Van't Hoff equation gives the shift of a chemical equilibrium with T: at higher T the shift is in the direction of positive standard enthalpy of reaction ΔH^0 i.e. in endothermic direction.
Similarly column-3 PR's of electrochemical systems are obtained by substitution of the column-3 μ_i-models into the electrochemical equilibrium condition derived in Sect. 2.2.7.

5.1.1 Closed Single-Reaction System

Here we consider a closed system prepared from a single dissociating compound k. At given intensive states of the coexistent phases n_k fixes the system completely: $f_{ext} = 1$. The other component mole numbers follow from the N-1 stoichiometric relations.

When we select the mole number n^β of one of the phases as the independent extensive of the system, we can formulate the generalised property differential $dE(T, n^\beta)$ as we did for a closed physical multi-phase system in Sect. 4.1.1.

The phase-transition coefficients attached to the $\omega - 1$ dependent n^α's are found by employing $\omega - 1$ of the $N - 1$ stoichiometric relations. The remaining $a = N - \omega$ equations provide additional relations between the x_i^α's of the system. This is illustrated for closed two-phase and three-phase systems prepared from ammonium-chloride and ammonium-bicarbonate.

$$1 \text{ mole } k \rightleftharpoons \sum_{\neq k} \nu_i \text{ mole } i$$

N − 1 stoichiometric relations

$$i \neq k: \frac{dn_i}{\nu_i} = \frac{dn_k}{-1}$$

Total property differential

$$\boxed{dE = \frac{\partial E}{\partial T} dT + \Delta\bar{E}\, dn^\beta}$$

$$\nu^\alpha \equiv \frac{\partial n^\alpha}{\partial n^\beta}, \quad \nu^\beta \equiv 1$$

$$\Delta\bar{E} \equiv \frac{\partial E}{\partial n^\beta} = \sum \nu^\alpha \bar{E}^\alpha$$

ω − 1 stoichiometric relations to find ω − 1 ν^α's

$$i \neq k: \quad d\left(\frac{n_i}{\nu_i} + n_k\right) = 0, \quad \frac{\partial}{\partial n^\beta}\left(\frac{n_i}{\nu_i} + n_k\right) = 0, \quad \sum_\alpha \left(\frac{x_i^\alpha}{\nu_i} + x_k^\alpha\right)\nu^\alpha = 0$$

$$x_k^\beta = 1: \boxed{\sum_{\neq\beta} \left(\frac{x_i^\alpha}{\nu_i} + x_k^\alpha\right)\nu^\alpha = -1}$$

a = N − ω stoichiometric relations between x_i^α's

$$j \neq i \neq k: \quad d\left(\frac{n_j}{\nu_j} - \frac{n_i}{\nu_i}\right) = 0, \quad \frac{\partial}{\partial n^\beta}\left(\frac{n_j}{\nu_j} - \frac{n_i}{\nu_i}\right) = 0, \quad \sum_\alpha \left(\frac{x_j^\alpha}{\nu_j} - \frac{x_i^\alpha}{\nu_i}\right)\nu^\alpha = 0$$

$$x_k^\beta = 1: \boxed{\sum_{\neq\beta} \left(\frac{x_j^\alpha}{\nu_j} - \frac{x_i^\alpha}{\nu_i}\right)\nu^\alpha = 0}$$

$NH_4Cl \rightleftharpoons NH_3 + HCl$	G–S G–L–S	a = 1 a = 0	$y_{HCl} = y_{NH_3}$
NH_4HCO_3 $\rightleftharpoons NH_3 + CO_2 + H_2O$	G–S G–L–S	a = 2 a = 1	$y_{H_2O} = y_{CO_2} = y_{NH_3}$ $\frac{x_{H_2O} - x_{NH_3}}{y_{H_2O} - y_{NH_3}} = \frac{x_{CO_2} - x_{NH_3}}{y_{CO_2} - y_{NH_3}}$

5.2.1 Closed-System FPR and Equivalents. Closed System with $f_{int} = f_{ext} = 1$. Gibbs Phase Rule

For the closed-system FPR alternative representations based on the definitions of H, F and G can again be formulated.

For a closed system with $f_{int} = f_{ext} = 1$, reversible heating connects dQ with $\Delta\bar{H}$ and $\Delta\bar{S}$. The free enthalpy of phase transition $\Delta\bar{G} = 0$. In the same way as for a non-reacting system the Clausius–Clapeyron equation can be found.

The Gibbs phase rule for the number of intensive independents of a multi-phase reaction system is obtained by considering the $2 + N$ properties T, p and N μ_i's as primary intensives. From this a number of constraining equations must be deducted: r chemical equilibrium conditions, one for each independent reaction, a additional relations between mole fractions and ω Gibbs-Duhem relations. Additional relations arise from the electroneutrality condition for an electrolyte solution or from stoichiometric relations as encountered in Sect. 5.1.1. For a closed system prepared from a single dissociating salt we find for the number of apparent components:

$$N' = N - r - a = N - 1 - (N - \omega) = \omega - 1.$$

The $\omega - 1$ non-solid phases can be regarded as independent "components" from which the dissociating salt and, hence, the closed system as a whole can be formed. The number of intensive degrees of freedom becomes:

$$f_{int} = 2 + N' - \omega = 2 + (\omega - 1) - \omega = 1.$$

Closed-system FPR and equivalents

$$U \qquad dU = TdS - pdV$$

$$H = U + pV \qquad dH = TdS + Vdp$$

$$F = U - TS \qquad dF = -SdT - pdV \qquad \left(\frac{\partial p}{\partial T}\right)_V = \left(\frac{\partial S}{\partial V}\right)_T$$

$$G = U + pV - TS \qquad dG = -SdT + Vdp$$

reversible heating: $dQ = dU + pdV = TdS$

Closed system with $f_{int} = f_{ext} = 1$

$$\Delta\bar{G} = \sum \frac{\partial n_i}{\partial n^\beta}\mu_i = -\sum \nu_i \mu_i \frac{\partial n_k}{\partial n^\beta} = 0$$

$$\Delta\bar{G} = \Delta\bar{H} - T\Delta\bar{S} = 0$$

$$(dQ)_p = \Delta\bar{H} dn^\beta = T\Delta\bar{S} dn^\beta, \quad \frac{dp}{dT} = \frac{\Delta\bar{H}}{T\Delta\bar{V}}$$

Gibbs phase rule

2 + N primary intensives: T, p, N μ_i's

r chemical equilibrium conditions: $\sum \nu_i \mu_i = 0$

a additional relations: electroneutrality condition,
stoichiometric relations

apparent number of components: $\boxed{N' = N - r - a}$

number of intensive degrees of freedom: $\boxed{f_{int} = 2 + N' - \omega}$

5.3.1 Chemical Equilibrium Constants

We shall now start to insert column-3 single-phase PR's in the chemical and electrochemical equilibrium conditions as derived in cell 2.2.

The resulting PR's involve standard properties. The standard free enthalpies of a compound were defined in Chapter 3. For a liquid-phase component they can be based on the zero-dilution state or the infinite-dilution state. For the infinite-dilution state as a basis the use of concentrations and molalities changes G_i^0 and its accompanying activity coefficient.

The standard free enthalpy of formation $\Delta G_{f,i}^0$ is by definition zero for an element and for H^+ (hydrogen-ion convention). The standard free enthalpy of reaction is composed of the values of $\Delta G_{f,i}^0$ of the components participating in the reaction.

Compression and mixing of the standard states of the reaction components to the conditions where chemical equilibrium prevails, yield a PR involving ΔG^0, ΔV_{cp}^0, the standard volume change of reaction for the condensed-phase components, non-ideality coefficients, T, p and composition parameters. Here we use for liquid-phase components the zero-dilution state as a basis.

The standard equilibrium constant K^0 increases with $-\Delta G^0$. The change of K^0 with T is given by the Van't Hoff equation. At higher T the equilibrium can be seen to shift in the direction of a positive ΔH^0, i.e. in the direction in which heat is absorbed, and vice versa.

In the column-3 equilibrium condition the stoichiometric coefficients of the reaction occur as exponents of the composition parameters. The equilibrium constant K is a weak function of p. Usually the effect of p of K is negligible and $K = K^0$.

phase	$G_i^o(T)$	composition parameter	non-ideality coefficient
solid	$\bar{G}_i\ (T, 0)$	—	—
liquid	$\bar{G}_i\ (T, 0)$	x_i	γ_i^z
	$\bar{G}_i^\infty (T, 0)$	x_i	γ_i^∞
	$\mu_i^\infty\ (T, 0, c_i = 1)$	c_i	$\gamma_i^{\infty, c}$
	$\mu_i^\infty\ (T, 0, l_i = 1)$	l_i	$\gamma_i^{\infty, l}$
gas	$\bar{G}_i^{ig}\ (T, 1)$	y_i	φ_i

standard free enthalpy of formation:	standard free enthalpy of reaction
$\Delta G_{f,i}^0 = G_i^0 - \sum_l b_{li} G_l^0$ $G_{f,l}^0 \equiv 0, \quad \Delta G_{f,H^+}^0 \equiv 0$	$\Delta G^0 = \sum_i \nu_i G_i^0 = \sum_i \nu_i \Delta G_{f,i}^0$

Equilibrium condition

$$\sum_i \nu_i \frac{\mu_i}{RT} = 0$$

$$\frac{\Delta G^0}{RT} + \frac{p\Delta V_{cp}^0}{RT} + \sum \{\nu_i \ln(\varphi_i y_i p) + \nu_j \ln(\gamma_j^z x_j)\} = 0$$

$$\boxed{\ln K^0 = -\frac{\Delta G^0}{RT}}$$ standard equilibrium constant

$$\boxed{\frac{d(\ln K^0)}{d\frac{1}{T}} = -\frac{\Delta H^0}{R}}$$ Van't Hoff equation

$$\boxed{\prod \{(\varphi_i y_i p)^{\nu_i} (\gamma_j^z x_j)^{\nu_j}\} = K^0 \exp\left(-\frac{p\Delta V_{cp}^0}{RT}\right) = K}$$

Application of the equilibrium constants to electrolyte solutions and to G–S, L–S and G–L systems is straightforward.

Electrolyte solutions

$H_2O \rightleftharpoons H^+ + OH^-$	$\ln K = \frac{\Delta G^0_{f,H_2O} - \Delta G^0_{f,OH^-}}{RT}$	$K = \frac{(\gamma^{\infty}_{H^+OH^-})^2[H^+][OH^-]}{\gamma^{z}_{H_2O} x_{H_2O}}$
$HCO_3^- \rightleftharpoons H^+ + CO_3^=$	$\ln K = \frac{\Delta G^0_{f,HCO_3^-} - \Delta G^0_{f,CO_3^=}}{RT}$	$K = \frac{(\gamma^{\infty}_{2H^+CO_3^=})^3[H^+][CO_3^=]}{(\gamma^{\infty}_{H^+,HCO_3^-})^2[HCO_3^-]}$
$HAc \rightleftharpoons H^+ + Ac^-$	$\ln K = \frac{\Delta G^0_{f,HAc} - \Delta G^0_{f,Ac^-}}{RT}$	$K = \frac{(\gamma^{\infty}_{H^+Ac^-})^2[H^+][Ac^-]}{\gamma^{\infty}_{HAc}[HAc]}$

Two-phase systems

G-S: $ZnS + \frac{3}{2}O_2 \rightleftharpoons ZnO + SO_2$	$\ln K = \frac{\Delta G^0_{f,ZnS} - \Delta G^0_{f,ZnO} - \Delta G^0_{f,SO_2}}{RT}$
	$K = \frac{\varphi_{SO_2} y_{SO_2} p}{(\varphi_{O_2} y_{O_2} p)^{3/2}}$
L-S: $K_{\nu^+}A_{\nu^-} \rightleftharpoons \nu_+ K^{+z+} + \nu_- A^{-z-}$	$\ln K = \frac{\Delta G^0_{f,K_{\nu^+}A_{\nu^-}} - \nu_+ \Delta G^0_{f,K_{\nu^+}} - \nu_- \Delta G^0_{f,A_{\nu^-}}}{RT}$
	$K = (\gamma^{\infty}_{\pm})^{\nu}[K^{+z_+}]^{\nu_+}[A^{-z_-}]^{\nu_-}$
G-L: $HCl \rightleftharpoons H^+ + Cl^-$	$\ln K = \frac{\Delta G^0_{f,HCl} - \Delta G^0_{f,Cl^-}}{RT}$
	$K = \frac{(\gamma^{\infty}_{H^+Cl^-})^2[H^+][Cl^-]}{\varphi_{HCl} y_{HCl} p}$
$CO_2 + 2NH_3 \rightleftharpoons CO(NH_2)_2 + H_2O$	$\ln K = \frac{\Delta G^0_{f,CO_2} + 2\Delta G^0_{f,NH_3} - \Delta G^0_{f,CO(NH_2)_2} - \Delta G^0_{f,H_2O}}{RT}$
	$K = \frac{\gamma^{\infty}_{CO(NH_2)_2} \gamma^{z}_{H_2O} [CO(NH_2)_2] x_{H_2O}}{(\varphi_{CO_2} y_{CO_2} p) \cdot (\varphi_{NH_3} y_{NH_3} p)^2}$

5.3.2 Electrochemical Systems

In addition to the column-2 PR's for the potential of an electrochemical system and for a single-electrode potential, standard potentials can be defined.

By inserting column-3 PR's for μ_i and neglecting the effect of p on μ_i for condensed-phase components, the potential e can be expressed as a function of e^0, composition parameters and non-ideality coefficients.

electrical potential $e = -\frac{\Delta\bar{G}}{\mathscr{F}}$

electrode potential $e_i = \frac{-\Delta\bar{G}_{Ox_i/Red_i}}{\mathscr{F}}, \quad e = e_1 - e_2$

standard potentials $e^0 = -\frac{\Delta G^0}{\mathscr{F}}$

$$e_i^0 = \frac{-\Delta G^0_{Ox_i/Red_i}}{\mathscr{F}}, \quad e^0 = e_1^0 - e_2^0$$

$$\begin{array}{ll} \tfrac{1}{2}Hg_2Cl_2 & \rightarrow Hg + Cl^- + 1\oplus \\ \tfrac{1}{2}H_2 + 1\oplus & \rightarrow H^+ \\ \hline \tfrac{1}{2}Hg_2Cl_2 + \tfrac{1}{2}H_2 & \rightarrow Hg + H^+ + Cl^- \end{array}$$

$$e^0_{Hg_2Cl_2/Hg} = \frac{\tfrac{1}{2}\Delta G^0_{f,\,Hg_2Cl_2} - \Delta G^0_{f,\,Cl^-}}{\mathscr{F}}$$

$$e^0_{H^+/H_2} = 0, \quad e^0 = e^0_{Hg_2Cl_2/Hg} - e^0_{H^+/H_2}$$

$$e = e^0 + \frac{RT}{\mathscr{F}} \ln \frac{(\bar{\varphi}_{H_2} p_{H_2})^{\frac{1}{2}}}{(\gamma^{\infty}_{H^+Cl^-})^2 [H^+][Cl^-]}$$

$$\begin{array}{ll} 2AgCl & \rightarrow 2Ag + 2Cl^- + 2\oplus \\ Zn + 2\oplus & \rightarrow Zn^{2+} \\ \hline \multicolumn{2}{l}{2AgCl + Zn \rightarrow 2Ag + Zn^{2+} + 2Cl^-} \end{array}$$

$$e^0 = \frac{2\Delta G^0_{f,\,AgCl} - \Delta G^0_{f,\,Zn^{2+}} - 2\Delta G^0_{f,\,Cl^-}}{2\mathscr{F}}$$

$$e = e^0 + \frac{RT}{2\mathscr{F}} \ln \frac{1}{(\gamma^{\infty}_{Zn^{2+},2Cl^-})^3 [Zn^{2+}][Cl^-]^2}$$

6 BE's of Non-Flow Systems

This is the first of a series of four chapters on BE's. It deals with non-flow systems. In practical terms one can think of a non-flow system as a piston-in-cylinder system with inlets and outlets. The intensive properties of the system are time-dependent and place-independent. Transport of matter into or from the system is in general not gradientless.
In deriving equations we use mole and mass balances and the FBE's [U] and [S]. Column-2 PR's are included to express physical equilibrium within the system. In this chapter column-3 equations are obtained by inserting ideal-gas PR's.

6.2 Column-2 Equations for Non-Flow Systems

First we derive a column-2 equation for δS_p from the BE's [n_i], [U] and [S] over a differential time interval, and the FPR. Contributions to δS_p which present themselves are due to departure from reversible expansion or compression, irreversible mixing at the inlets and to irreversible chemical reaction within the system. Each of the contributions was found separately in suitably devised systems in cell 2.2.

A non-flow system can pass through cycles. In the BE's over a cycle the accumulation terms vanish as all properties at the end of a cycle have returned to their initial values. Reversible closed cycles, adiabatic cycles and isothermal cycles are distinguished.

In a Carnot cycle the system cycles reversibly through two isotherms and two adiabates (isentropes). In the heat engine mode heat cascades from a higher to a lower temperature and is partly converted into work. In the heat pump mode the energy transports are reversed. With the aid of the column-2 equation $Q_1/Q_2^- = T_1/T_2$ an absolute temperature scale can be devised.

For reversible heating/compression cycles of an open piston-in-cylinder system we derive equations connecting W and Q with property changes.

For the sake of curiosity we devise idealised systems which accomplish the various cases of interface transport described in Sect. 2.2.4 reversibly. The reversibly operating devices employed are membranes for the selective transport of matter, pistons which gain work from the displacement of the interface, Carnot heat engines and heat pumps to transport, heat, and open piston-in-cylinder systems for heating/cooling and compression/expansion. In each case the net work output of the devices equals the exergy loss found in Sect. 2.2.4 in its irreversible counterpart.

To describe a batch reactor with heating or cooling by a coil, we apply the BE's [m], [m_i] and [U] per unit time. By inserting the PR $d\bar{h}(T, p, w_i)$ the heat balance emerges.

6.3 Column-3 Equations for Non-Flow Systems

Here we use findings in cell 6.2 in combination with ideal-gas PR's presented in cell 3.3.

First we derive equations for heating/cooling and compression/expansion of a closed system. The system is described by the balances [U] and [S] and the column-2 equation for δS_p. Column-2 equations are obtained for Q and W or S_p for various paths: isochoric, isobaric, adiabatic (isentropic) and isothermal reversible paths, and adiabatic and isothermal irreversible paths. By substituting ideal-gas PR's these equations contain only measurable properties such as T, p and V at the initial and final states.

The derivation of equations for an irreversible reaction in a closed system proceeds similarly. Equations for the rise in T and S_p for an isochoric (constant U) and an isobaric (isenthalpic) adiabatic path are presented.

The remainder of the cell deals with cycles composed of connected trajects of constant S, V, p and T. In general column-3 equations are obtained for W^-/Q for a cycle in the heat engine mode and W/Q in the refrigeration mode.

6.2.1 BE's Over a Time Interval. Charging/Discharging

In this chapter we consider idealised non-flow systems. In practical terms a non-flow system could comprise the volume enclosed by a piston in a cylinder. Within the system intraphase and interface physical equilibria are assumed to be established at all times. In other words, all intensive properties are gradientless within each phase of the system, while T, p and the μ_i's are gradientless throughout the system. Inputs of matter, work and heat in addition to chemical reactions within the system cause the system properties to change with time.

From the FBE's [n_i], [U] and [S] and the FPR of matter which reflects that within the system physical equilibrium prevails at all times, an expression is found for $T\delta S_p$. The expression reflects departure from reversible expansion or compression, irreversible mixing at the inlet and irreversible chemical reaction. The reversible expansion work $dW^- = pdV$ or compression work $dW = -pdV$ can be represented in a p-V diagram. The contribution of inlet mixing to $T\delta S_p$ corresponds with the entropy production for irreversible interface transport of matter as found in Sect. 2.2.4. This contribution is zero when all intensive properties are gradientless across inlet and outlets. The term for an irreversible chemical reaction is as encountered previously.

We now consider reversible adiabatic charging/discharging of a non-flow system. Application of the BE's [n] and [S] reveals that $d\bar{S} = 0$. For isobaric charging/discharging all intensive properties remain constant with time, whereas for isochoric discharging only $\bar{S}$ remains constant.

BE's over a time interval

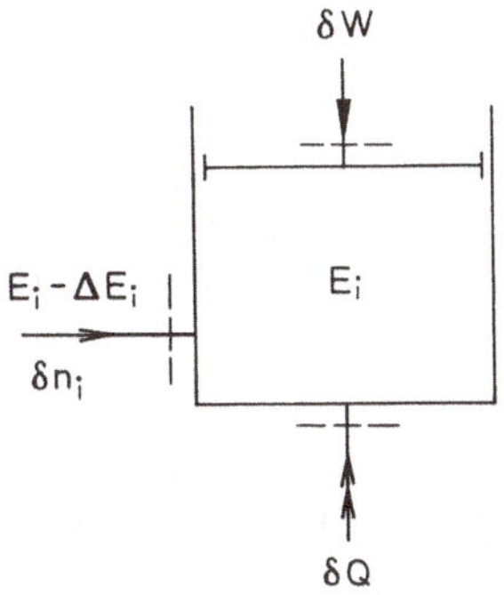

$[n_i]$: $dn_i = \delta n_i + \nu_i \delta n_{kp}$

$[U]$: $dU = \delta H + \delta Q + \delta W$

$[S]$: $dS = \delta S + \frac{\delta Q}{T} + \delta S_p$

FPR: $dU = TdS - pdV + \sum \mu_i dn_i$

$[U] + T[S] - \sum \mu_i [n_i]$:

$$-pdV = \sum \{(H_i - \Delta H_i) - T(S_i - \Delta S_i) - \mu_i\} \delta n_i + \delta W - T\delta S_p - \sum \nu_i \mu_i \delta n_{kp}$$

$$\boxed{T\delta S_p = (pdV - \delta W^-) + T\sum\left(\Delta S_i - \frac{\Delta H_i}{T}\right)\delta n_i - \sum \nu_i \mu_i \delta n_{kp}}$$

reversible expansion $\boxed{dW^- = pdV}$

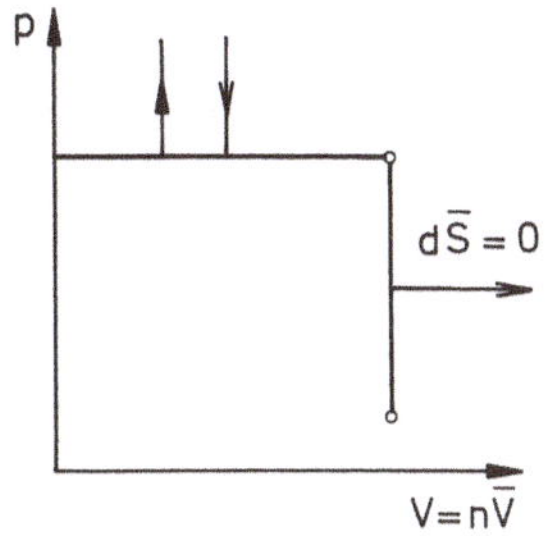

Charging/discharging

$[n]$: $dn = \delta n$

$[S]$: $d(n\bar{S}) = \bar{S}\delta n \quad d\bar{S} = 0$

isobaric charging/discharging: $dp = 0$, all intensives constant

isochoric discharging: $d(n\bar{V}) = 0 \quad d\bar{S} = 0$

6.2.2 Cycles

For a cycling non-flow system all intensive and extensive properties change periodically with time. In the BE's over the cycle time t_c the accumulation terms vanish as the properties at the end of a cycle have returned to their initial values.
When compression/expansion is reversible throughout the cycle, the net work output W^- during a cycle is given by a contour integral in a p-V diagram.
For a reversible closed cycle the net transport of U and S by matter into the system becomes zero. The net work output W^- equals the net heat input Q during a cycle. W^- and Q are given by the contour integrals in the p-V and T-S diagrams of the cycling system.
For an adiabatic cycle δQ and Q are zero.
For an isothermal cycle, finally, [U] and [S] can be combined to an expression in which Q has been eliminated and the net transport of G by matter occurs.

[U]: $$0 = \oint \delta H + Q + W$$

[S]: $$0 = \oint \delta S + \oint \frac{\delta Q}{T} + S_p$$

Reversible compression/expansion

$$dW = -\,pdV, \quad W = -\oint pdV = -\oint d(pV) + \oint Vdp = \oint Vdp$$

Reversible heating of closed system

$$dQ = TdS$$

Reversible closed cycle

[U]: $$0 = Q + W$$

[S]: $$0 = \oint \frac{dQ}{T}$$

$$W^- = Q = \oint pdV = \oint TdS$$

Adiabatic cycle

[U]: $$0 = \oint \delta H + W$$

[S]: $$0 = \oint \delta S + S_p$$

Isothermal cycle

[U]: $$0 = \oint \delta H + Q + W$$

[S]: $$0 = \oint \delta S + \frac{Q}{T} + S_p$$

$$0 = \oint \delta G + W - T\delta S_p$$

6.2.3 Carnot Cycle

A Carnot cycle is a reversible closed cycle comprising two isotherms and two isentropes.
Application of the FBE's [U] and [S] reveals that the quantity of heat Q_1 cascading from a higher temperature T_1 to a lower temperature T_2 is partly converted into work. The work output W^- of a Carnot engine can either be expressed in Q_1, the heat absorbed from a hot reservoir or in Q_2^-, the heat rejected to a cold heat sink.
Similar expressions can be derived for the work W required to "pump" a quantity of heat Q_1^- to a hot heat reservoir or to "pump" a quantity of heat Q_2 from a cold reservoir.

The heat quantities Q_1 and Q_2^- have the same ratio as the temperatures T_1 and T_2. In Sect. 1.2.1 we noted that the equivalence of heat and work as energy being transported defines heat in terms of work. Now we can illustrate the existence of an absolute temperature scale when a value is assigned to T at a single specified temperature. The Kelvin scale is based on $T_2 \approx 273$ K for melting ice at atmospheric pressure. The temperature of any system can be determined by applying a Carnot cycle between that system and melting ice at atmospheric pressure and measuring the ratio of the heat quantities. For boiling water at atmospheric pressure, for example, it is found that $T_1 = Q_1/Q_2^- \cdot T_2 \approx 373$ K.

Engines

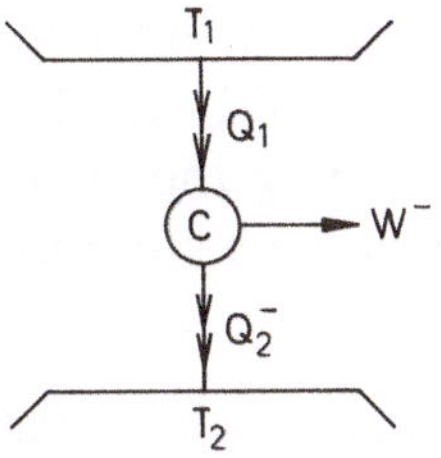

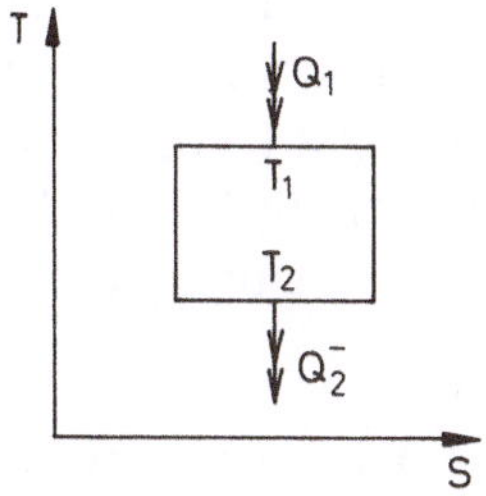

[U]: $0 = Q_1 - Q_2^- - W^-$

[S]: $0 = \frac{Q_1}{T_1} - \frac{Q_2^-}{T_2}$

$$\boxed{\frac{Q_1}{Q_2^-} = \frac{T_1}{T_2}}, \quad \boxed{W^- = Q_1 - Q_2^- = Q_1\left(1 - \frac{T_2}{T_1}\right) = Q_2^-\left(\frac{T_1}{T_2} - 1\right)}$$

heat engine — engine on cold heat sink

Heat pumps

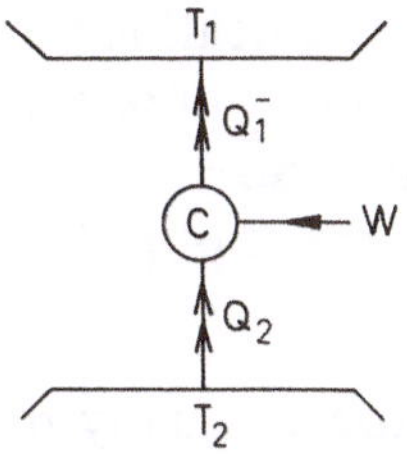

[U]: $0 = - Q_1^- + Q_2 + W$

[S]: $0 = - \frac{Q_1^-}{T_1} + \frac{Q_2}{T_2}$

$$\boxed{W = Q_1^- - Q_2 = Q_1^-\left(1 - \frac{T_2}{T_1}\right) = Q_2\left(\frac{T_1}{T_2} - 1\right)}$$

heat pump — refrigerator

6.2.4 Heating/Compression Cycles

Consider a reversible open cycle of a piston-in-cylinder system. The cycle comprises: isobaric charging at constant intensive properties, heating/compression of the charge to change the intensives from inlet to outlet values, isobaric discharging at constant intensives and, finally, return to the inlet pressure by closing the outlet and opening the inlet. The net work input W by the piston during a cycle is given by a contour integral in the p-V diagram.

The property changes $\Delta\bar{H}$ and $\Delta\bar{S}$ during the closed traject of the cycle follow from the expression for reversible heating and the FPR of a closed system.

The FBE's [U] and [S] of the piston-in-cylinder system over a cycle yield the same relations. In the exergy balance of the system the exergy input by heat can be interpreted as the total Carnot work W_c of an infinite number of cycles of a Carnot heat pump. During a cycle of the Carnot heat pump the quantity of heat dQ is delivered at the instantaneous value of T of the piston-in-cylinder system. The total work input $W + W_c$ equals the exergy increase $\Delta\bar{\varepsilon}.n$ of a charge. This result could have been obtained directly by applying an exergy balance to the combined system of the piston-in-cylinder system and the Carnot heat pump.

A number of special cases can be distinguished. For isobaric heating the contour integral in the p-V diagram and therefore the net input of piston work W is zero. For adiabatic (isentropic) compression $\Delta\bar{S} = 0$ and $W = \Delta\bar{H}n$. For isothermal compression the required net input of piston work $W = \Delta\bar{G}n$.

The relations presented hold for a single phase. They apply equally to a multi-phase system when replacing the average molar single-phase properties $\bar{E}$ by the average molar multi-phase properties $\tilde{E}$.

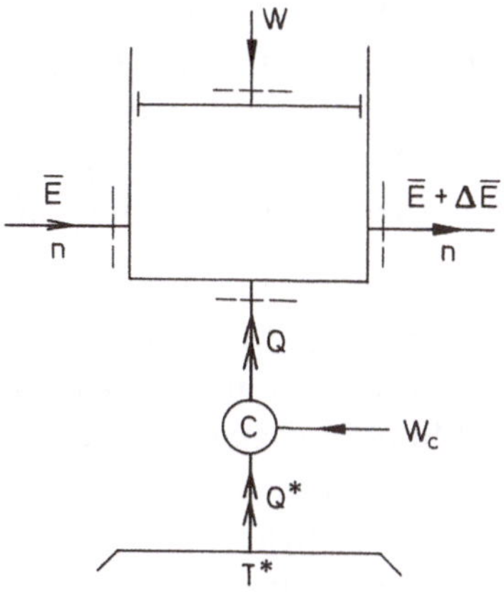

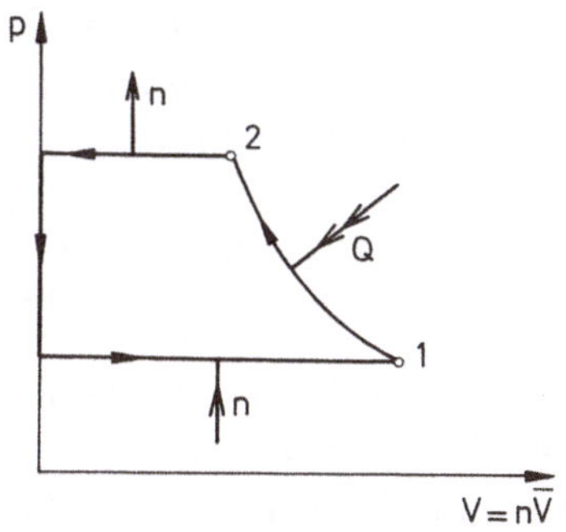

$$W = -\oint p\,dV = \oint V\,dp = \int_1^2 V\,dp$$

Property changes during closed traject

reversible heating: $dQ = TdS$, FPR: $dH = TdS + Vdp$

$$\Delta\bar{H}n = \int_1^2 (TdS + Vdp) = Q + W \qquad \Delta\bar{S}n = \int_1^2 \frac{dQ}{T}$$

FBE's of piston-in-cylinder system over a cycle

[U]: $0 = -\Delta\bar{H}n + Q + W$

[S]: $0 = -\Delta\bar{S}n + \int_1^2 \frac{dQ}{T}$

[U] − T*[S]: $0 = -\Delta\bar{\varepsilon}n + \int_1^2 dQ\left(1 - \frac{T^*}{T}\right) + W$ $\quad W_c = \int_1^2 dQ\left(1 - \frac{T^*}{T}\right)$

$$W + W_c = \Delta\bar{\varepsilon}n = (\Delta\bar{H} - T^*\Delta\bar{S})n$$

Special case	Q	W	W_c
isobaric heating	$Q = \Delta\bar{H}n$	$W = 0$	$W_c = \Delta\bar{\varepsilon}n$
adiabatic compression	$\Delta\bar{S} = 0$	$W = \Delta\bar{H}n$	$W_c = 0$
isothermal compression	$Q = T\Delta\bar{S}n$	$W = \Delta\bar{G}n$	$W_c = (T - T^*)\Delta\bar{S}n$

6.2.5 Reversible Interface Transport

Starting from the expressions for the transport of U and S, δU_x and δS_x as defined in a single-phase at local equilibrium, we investigated in Sect. 2.2.4 irreversible interface transport of matter and heat, and displacement of an interface. By applying the FBE's [U] and [S] it was found that δU_x is continuous when crossing the interface, while δS_x exhibits a step increase equal to δS_p.

How can we now devise an idealised system for reversible rather than irreversible transport between the two phases at given intensive states of the phases and given δn_i, δQ_1, δQ_2 and δV_x? Such a system involves:

- a piston in a cylinder for reversible displacement of the interface δV_x.
- for each component a combined system of a piston-in-cylinder system and a Carnot heat pump for a heating/decompression cycle of a charge δn_i. Supply and delivery of the charge is through membranes with reversible non-isothermal transport (in 2.2.4 we found such transport to be isenthalpic and isentropic).
- a Carnot heat engine withdrawing δQ_1 from phase 1 and a Carnot heat pump delivering δQ_2 to phase 2.

From the FBE's [U] and [S] of the total interface system it follows that the exergy loss $T^*\delta S_p$ accompanying irreversible interface transport, is recovered in the reversibly operating system by conversion of a net heat input dQ^* from an infinite reservoir with temperature T^* into a net work output dW^-. It can be readily verified that dW^- equals the net work output of the reversibly operating devices.

Transport in a single phase at equilibrium

$$\delta U_x = \sum H_i \delta n_i + \delta Q - p\delta V_x, \quad \delta S_x = \sum S_i \delta n_i + \frac{\delta Q}{T}$$

Irreversible interface transport

$$[U]: \quad 0 = -\Delta(\delta U_x) \qquad\qquad \boxed{\Delta(\delta U_x) = 0}$$

$$[S]: \quad 0 = -\Delta(\delta S_x) + \delta S_p, \qquad\qquad \boxed{\delta S_p = \Delta(\delta S_x)}$$

Interface system for reversible transport

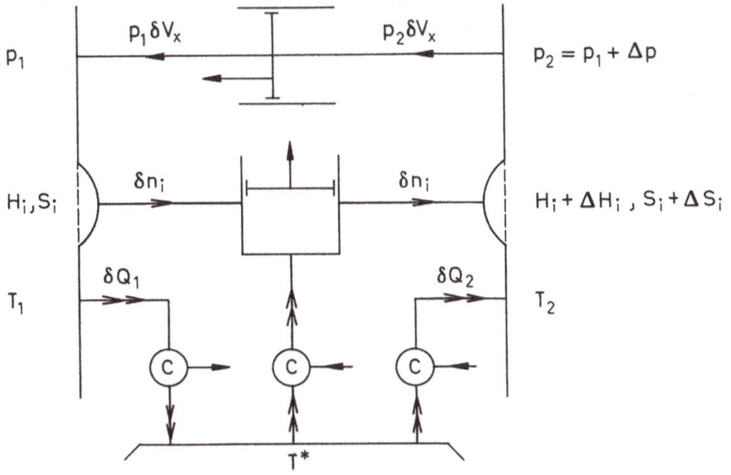

$$[U]: \quad 0 = -\Delta(\delta U_x) + dQ^* - dW^-$$

$$[S]: \quad 0 = -\Delta(\delta S_x) + \frac{dQ^*}{T^*} \qquad\qquad \boxed{dW^- = dQ^* = T^*\Delta(\delta S_x)}$$

$$dW^- = \delta V_x \Delta p - \sum(\Delta H_i - T^*\Delta S_i)\delta n_i$$
$$+ \delta Q_1\left(1 - \frac{T^*}{T_1}\right) - \delta Q_2\left(1 - \frac{T^*}{T_2}\right)$$
$$= -\Delta\{\sum H_i \delta n_i + \delta Q - p\delta V_x\} + T^*\Delta\left\{\sum S_i \delta n_i + \frac{\delta Q}{T}\right\}$$
$$= -\Delta(\delta U_x) + T^*\Delta(\delta S_x) = T^*\Delta(\delta S_x)$$

In Sect. 2.2.4 we discussed a number of special cases of irreversible interface transport. Their reversible counterparts with recovery of exergy loss as a net work output of reversibly operating devices are described here.

Reversible transport of heat δQ can be accomplished by a Carnot heat engine and a Carnot heat pump.

Reversible non-isothermal transport of matter δn_i involves in addition to a membrane/piston-in-cylinder with Carnot heat pump/membrane assembly an additional Carnot heat pump delivering the heat δQ_2 accompanying irreversible non-isothermal transport of matter.

Reversible interface displacement is accomplished by a piston separating the two phases. Heat from dissipation of work in the irreversible case is here matched by heat delivered by a Carnot heat pump.

Reversible transport of heat

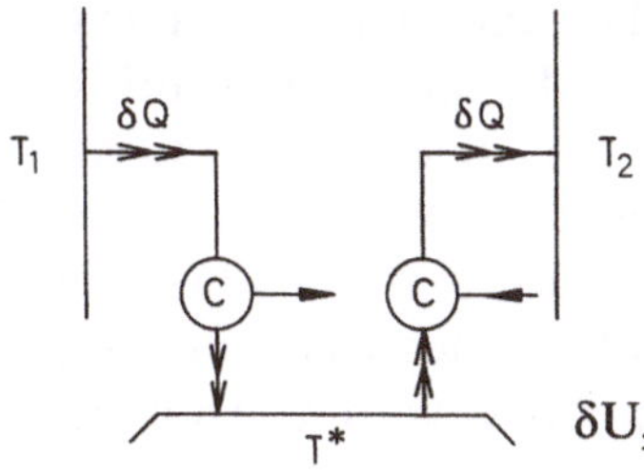

$$\delta U_x = \delta Q_1 = \delta Q_2 = \delta Q,$$

$$dW^- = \delta Q\left(1 - \frac{T^*}{T_1}\right) - \delta Q\left(1 - \frac{T^*}{T_2}\right) = T^*\delta Q \Delta\left(\frac{1}{T}\right)$$

Reversible non-isothermal transport of matter

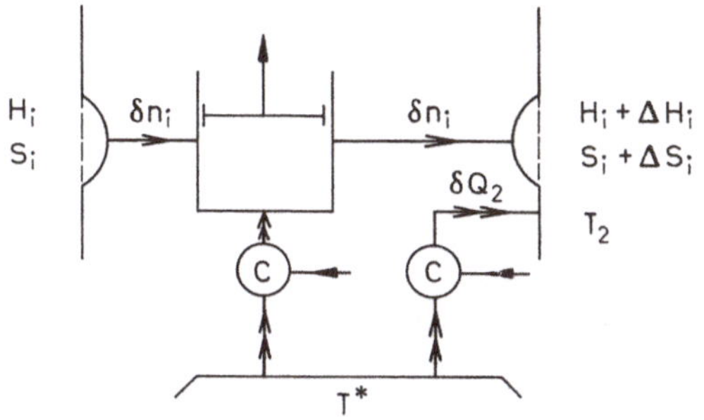

$$\delta U_x = \sum H_i \delta n_i = \sum (H_i + \Delta H_i)\delta n_i + \delta Q_2, \quad \delta Q_2 = -\sum \Delta H_i \delta n_i$$

$$dW^- = -\sum(\Delta H_i - T^*\Delta S_i)\delta n_i - \left(-\sum \Delta H_i \delta n_i\right)\left(1 - \frac{T^*}{T_2}\right)$$

$$= T^*\sum\left(\Delta S_i - \frac{\Delta H_i}{T_2}\right)\delta n_i$$

Reversible displacement

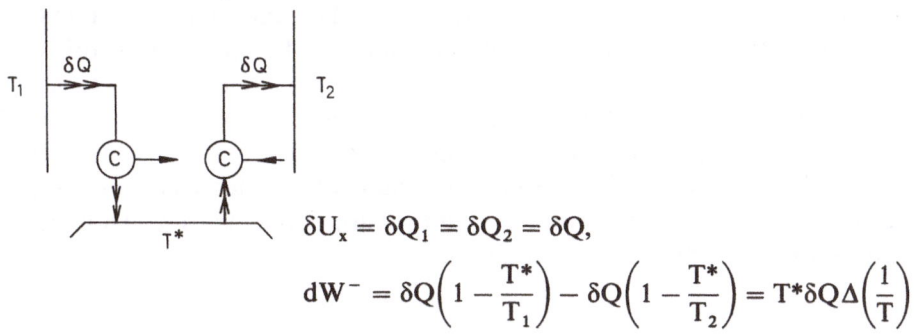

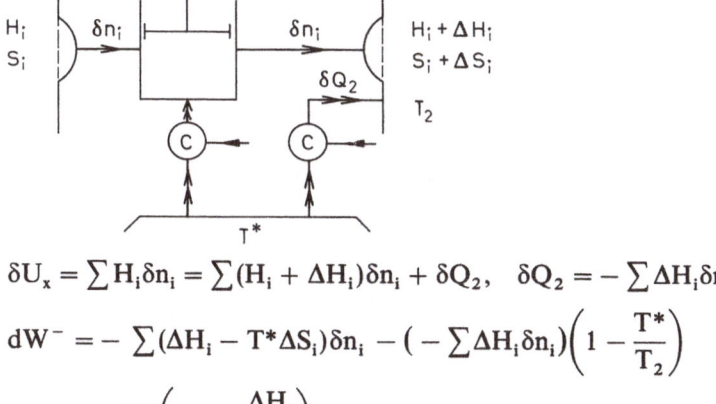

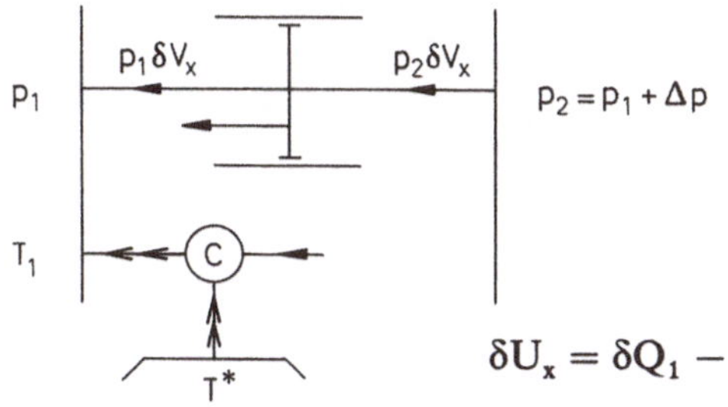

$$\delta U_x = \delta Q_1 - p_1 \delta V_x = -p_2 \delta V_x, \quad \delta Q_1 = -\Delta p \delta V_x$$

$$dW^- = \Delta p \delta V_x - \Delta p \delta V_x\left(1 - \frac{T^*}{T_1}\right) = T^*\frac{\Delta p \delta V_x}{T_1}$$

Reversible isothermal transport of matter involves reversible isothermal membrane transport with gradientless μ_i. The combined heat effect accompanying the transport through the left-hand membranes is utilised in a Carnot heat engine. The values of H_i in front of the right-hand membranes are kept on the values of H_i in phase 1 to ensure that the heat effect of irreversible isothermal transport of matter is matched. The second component of dW^- equals the sum of the decreases in exergy accompanying the isothermal decompression cycle with Carnot heat supply for the individual components.

Reversible isothermal transport of matter

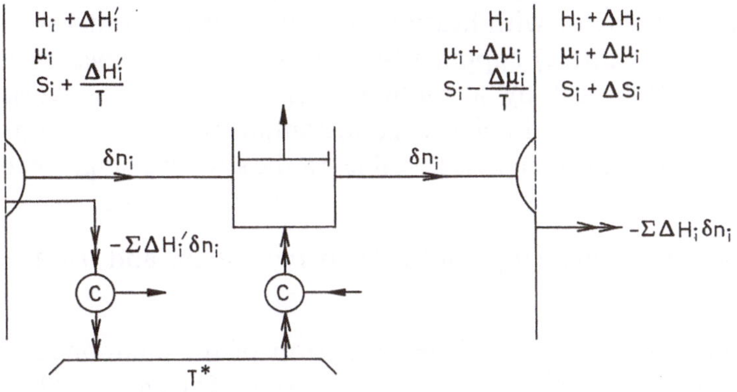

$$dW^- = \left(-\sum \Delta H_i' \delta n_i\right)\left(1 - \frac{T^*}{T}\right) + \sum \left\{\Delta H_i' - T^*\left(\frac{\Delta H_i'}{T} + \frac{\Delta \mu_i}{T}\right)\right\} \delta n_i$$

$$= -T^* \sum \frac{\Delta \mu_i}{T} \delta n_i$$

6.2.6 Closed-System BE's per Unit Time

Having described reversible interface transport with idealised reversibly operating devices, we shall now focus on the more practical problem of a batch chemical reactor with heating or cooling. We consider a closed system with a heat rate $\dot{Q}$ supplied by a heating coil (a negative value implies heat withdrawal by a cooling coil) and a rate of work $\dot{W}'$ dissipated by a mechanical agitator as instantaneous inputs. Compression or expansion of the system is possible. In the system a single reaction proceeds.

We apply the BE's [m], [m_i] and [U] to the system and use the PR $d\bar{h}(T, p, w_i)$.

The total mass m is constant. The final form of the component mass balance [m_i] describes the change of w_i with time. The chemical kinetics denoted as $\dot{n}_{kp}/m$, the molar rate of conversion to key component k per unit mass, are a function of the instantaneous intensive state only and can be determined experimentally.

In the internal-energy balance [U], $\bar{u}$ can be seen to increase with time by heat supply, irreversible conversion of work by the agitator in U and reversible conversion of compression work in U.

The enthalpy balance [H] can be interpreted similarly.

Combination of [H], the component mass balances [m_i] and the PR $d\bar{h}(T, p, w_i)$ results in what we refer to as the heat balance. It is noted that this terminology suggests incorrectly that heat can be regarded as an extensive property representing a form of energy stored rather than a form of energy being transported. Nevertheless we shall adhere to this widely applied terminology. In the heat balance the temperature can be seen to rise by supply of heat, dissipation of mechanical work, heat of reaction, and reversible compression. The last term vanishes when the thermal expansion coefficient is zero, which is usually to a good approximation true for a liquid mixture, or when the operation in the batch reactor is isobaric.

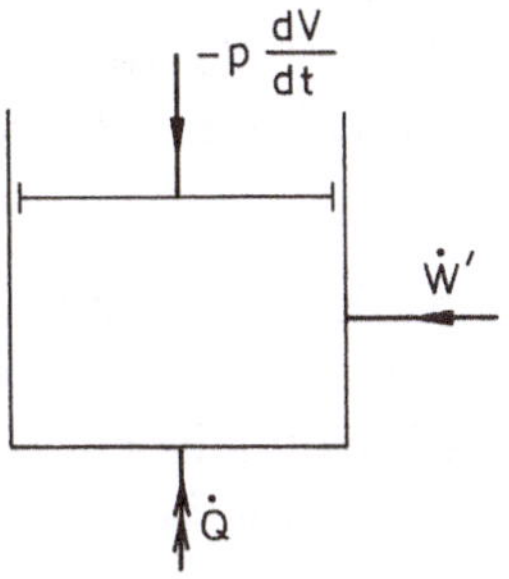

[m]: $\frac{dm}{dt} = 0, \quad m = \text{constant}$

$$[m_i]: \quad \boxed{\frac{dw_i}{dt} = \nu_i M_i \frac{\dot{n}_{kp}}{m}}$$

$$[U]: \quad \frac{d\bar{u}}{dt} = \frac{\dot{Q} + \dot{W}'}{m} + \left(-p\frac{d\bar{v}}{dt}\right)$$

$$PR: \quad d\bar{h} = \bar{c}_p dT + \left(\bar{v} - T\frac{\partial \bar{v}}{\partial T}\right)dp + \sum h_i dw_i$$

Enthalpy balance

$$\bar{h} = \bar{u} + p\bar{v}$$

$$[H]: \quad \boxed{\frac{d\bar{h}}{dt} = \frac{\dot{Q} + \dot{W}'}{m} + \bar{v}\frac{dp}{dt}}$$

Heat balance

$[H] - \sum h_i [m_i]$:

$$\bar{c}_p \frac{dT}{dt} + \left(\bar{v} - T\frac{\partial \bar{v}}{\partial T}\right)\frac{dp}{dt} = \frac{\dot{Q} + \dot{W}'}{m} + \bar{v}\frac{dp}{dt} - \sum \nu_i H_i \frac{\dot{n}_{kp}}{m}$$

$$\boxed{\bar{c}_p \frac{dT}{dt} = \frac{\dot{Q} + \dot{W}'}{m} + \left(-\sum \nu_i H_i\right)\frac{\dot{n}_{kp}}{m} + T\left(\frac{\partial \bar{v}}{\partial T}\right)_p \frac{dp}{dt}}$$

6.3.1 Heating/Cooling and Compression/Expansion of a Closed System

In this section we shall apply the FBE's and the expression for $T\delta S_p$ as derived in Sect. 6.2.1 to a closed non-flow system. Simultaneously we shall insert ideal-gas PR's as found in Sect. 3.3.5 to obtain column-3 relations.

For reversible processes $\delta S_p = 0$ and the compression work $dW = -pdV$.
The expressions for isochoric and isobaric heating are as found before. For an ideal gas the heat capacities are here assumed to be independent of T.
Reversible adiabatic compression is isentropic. Compression is accompanied by a rise in T, expansion by a drop in T. The compression work $W = \Delta U$. For an ideal gas the effect of T on heat capacities is again neglected.
For reversible isothermal compression of an ideal gas $\Delta U = 0$. The compression work is rejected as heat.

For irreversible expansion of a closed system the work output $\delta W^- = 0$. The loss of potential expansion work is reflected in the expression $T\delta S_p = pdV$.
Adiabatic irreversible expansion of a gas can be realised by expansion of that gas present in a volume into an evacuated second volume. When going from the initial state where the two volumes are separated from each other, to the final state $Q = 0$, $W^- = 0$ and $\Delta U = 0$. The accompanying temperature effect, the Joule effect, is zero for an ideal gas.
For an ideal gas adiabatic and isothermal irreversible expansion coincide.

[U]: $dU = \delta Q + \delta W$

[S]: $dS = \frac{\delta Q}{T} + \delta S_p$

$T\delta S_p = pdV - \delta W^-$

$\delta S_p=0$, $dW=-pdV$ reversible	$dQ=dU+pdV=TdS$		$dW=-pdV=dU-TdS$	
		ideal gas		ideal gas
isochoric heating	$dQ=dU$	$Q=n\bar{C}_v(T_2-T_1)$	$dW^-=0$	
isobaric heating	$dQ=dH$	$Q=n\bar{C}_p(T_2-T_1)$	$dW^-=pdV$	$W^-=nR(T_2-T_1)$
adiabatic (isentropic) compression	$dS=0$	$TV^{\kappa-1}=\text{constant}$	$dW=dU$	$W=n\bar{C}_v(T_2-T_1)$ $=\frac{nRT_1}{\kappa-1}\times\left\{\left(\frac{V_1}{V_2}\right)^{\kappa-1}-1\right\}$
isothermal compression	$dQ=TdS$	$Q=-nRT\ln\frac{V_1}{V_2}$ $=-nRT\ln\frac{p_2}{p_1}$	$dW=dF$	$W=nRT\ln\frac{V_1}{V_2}$ $=nRT\ln\frac{p_2}{p_1}$
$\delta W^-=0$, $T\delta S_p=pdV$ irreversible expansion	$\delta Q=dU=TdS-pdV$		$T\delta S_p=pdV=TdS-dU$	
		ideal gas		ideal gas
adiabatic expansion	$dU=0$	$dT=0$	$\delta S_p=dS$	$S_p=nR\ln\frac{V_2}{V_1}=$ $=nR\ln\frac{p_1}{p_2}$
isothermal expansion	$\delta Q=dU$	$\delta Q=0$	$T\delta S_p=-dF$	$T\delta S_p=nRT\ln\frac{V_2}{V_1}$ $=nRT\ln\frac{p_1}{p_2}$

6.3.2 Irreversible Reaction in a Closed System

Along the same lines as in the previous section we consider here an irreversible reaction proceeding in a closed non-flow system. Entropy production is entirely due to the reaction. Expansion is reversible: $\delta W^- = pdV$.

For an adiabatic irreversible reaction $S_p = \Delta S$. Two cases are considered: an isochoric and an isobaric adiabatic irreversible reaction with $\Delta U = 0$ and $\Delta H = 0$, respectively.

For an ideal-gas mixture the property changes ΔU, ΔH and ΔS are readily found starting from the standard properties of the reactants and inerts at the initial temperature T_1. From $\Delta U = 0$ and $\Delta H = 0$ the final temperature T_2 is found. The expression for S_p can be used to estimate the exergy loss in an internal-combustion engine.

[U]: $dU = \delta Q - \delta W^-$

[S]: $dS = \dfrac{\delta Q}{T} + \delta S_p$

$$T\delta S_p = (pdV - \delta W^-) - \sum \nu_i \mu_i \delta n_{kp}$$

irreversible reaction	$T\delta S_p = -\sum \nu_i \mu_i \delta n_{kp}$ $\delta W^- = pdV$	$\delta Q = dU + pdV$
adiabatic isochoric isobaric	$\delta S_p = dS$	$dU + pdV = 0$ $dU = 0$ $dH = 0$

Ideal gas

isochoric

$$\boxed{\Delta U = \sum \nu_i U_i^0(T_1) n_{kp} + n_2 \bar{C}_{v,2}(T_2 - T_1) = 0} \qquad V = n_1 \frac{RT_1}{p_1} = n_2 \frac{RT_2}{p_2}$$

isobaric

$$\boxed{\Delta H = \sum \nu_i H_i^0(T_1) n_{kp} + n_2 \bar{C}_{p,2}(T_2 - T_1) = 0}$$

$$V_1 = \frac{n_1 RT_1}{p}, \; V_2 = n_2 \frac{RT_2}{p}$$

$$S_p = \Delta S = \sum n_{i,2} \left\{ S_i^0(T_1) + C_{p,i}^0 \ln \frac{T_2}{T_1} - R \ln y_{i,2} p_2 \right\} - \sum n_{i,1} \left\{ S_i^0(T_1) \qquad - R \ln y_{i,1} p_1 \right\}$$

$$\boxed{\begin{aligned} S_p = \sum \nu_i S_i^0(T_1) n_{kp} + n_2 \bar{C}_{p,2} \ln \frac{T_2}{T_1} - (n_2 R \ln p_2 - n_1 R \ln p_1) \\ - (n_2 R \sum y_{i,2} \ln y_{i,2} - n_1 R \sum y_{i,1} \ln y_{i,1}) \end{aligned}}$$

6.3.3 Otto and Diesel Cycles

In addition to the Carnot cycle passing through connected trajects of constant S, T, S and T other reversible closed cycles of a simple type present themselves:

— cycle of constant S-V-S-V (Otto cycle)
— cycle of constant S-p-S-V (Diesel cycle)
— cycle of constant S-p-S-p (Brayton cycle)
— cycle of constant T-V-T-V (Stirling cycle)
— cycle of constant T-p-T-p (Ericsson cycle)

We shall explore these cycles for an ideal gas. Work/heat ratios for heat engine and refrigeration modes will be derived. In this section we start with the Otto and Diesel cycles.

For an Otto cycle [U] over a cycle combined with the expressions for isochoric heating and cooling and the PR's for isentropic compression and expansion, yields the efficiency W^-/Q. The higher the volume compression ratio φ, the higher the efficiency.
For an open Otto cycle isochoric adiabatic combustion ($dU = 0$) and isochoric discharge ($d\bar{S} = 0$) replace isochoric heating and cooling. The cycle starts with isobaric charging and ends with isobaric discharging.

The expression for W^-/Q for a Diesel cycle is obtained similarly. For a closed cycle isobaric heating replaces isochoric heating, while for an open cycle isobaric adiabatic combustion ($dH = 0$) comes in place of isochoric adiabatic combustion ($dU = 0$).

Otto cycle

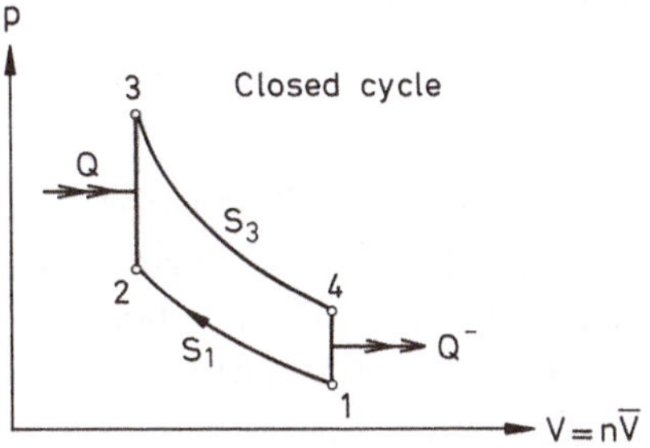

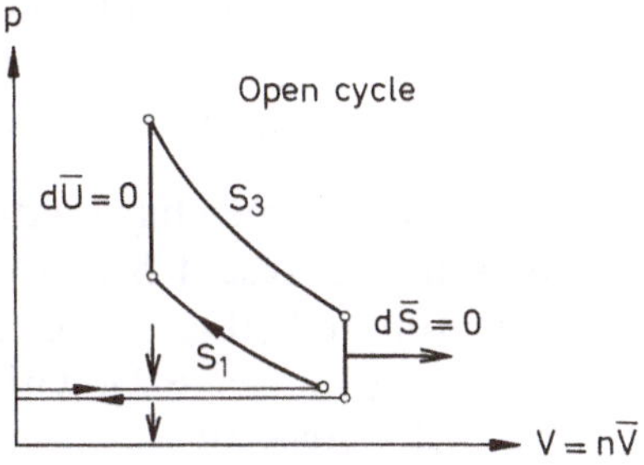

$$[U]: \quad 0 = Q - Q^- - W^-, \quad \frac{W^-}{Q} = 1 - \frac{Q^-}{Q}$$

heating/cooling: $\dfrac{Q^-}{Q} = \dfrac{T_4 - T_1}{T_3 - T_2} = \dfrac{T_1}{T_2}\dfrac{T_4/T_1 - 1}{T_3/T_2 - 1}$

PR's: $\dfrac{T_2}{T_1} = \dfrac{T_3}{T_4} = \varphi^{\kappa - 1} \quad \varphi = \dfrac{V_1}{V_2}$

$$\boxed{\frac{W^-}{Q} = 1 - \frac{T_1}{T_2} = 1 - \frac{T_3}{T_4} = 1 - \varphi^{-(\kappa - 1)}}$$

Diesel cycle

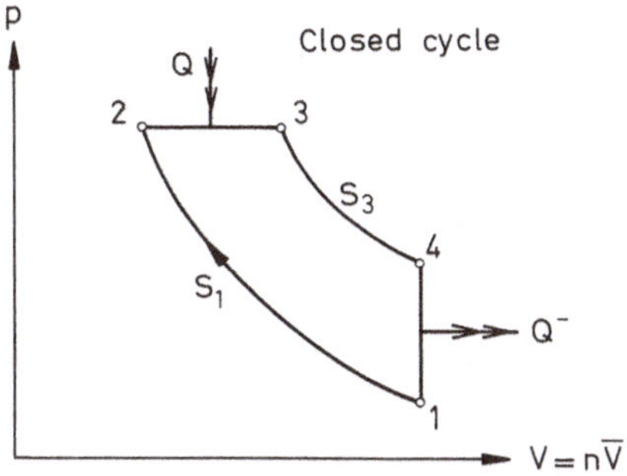

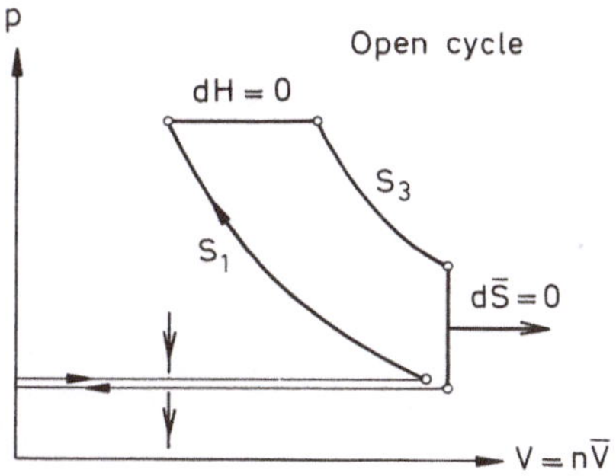

$$[U]: \quad 0 = Q - Q^- - W^-, \quad \frac{W^-}{Q} = 1 - \frac{Q^-}{Q}$$

heating/cooling: $\dfrac{Q^-}{Q} = \dfrac{1}{\kappa}\dfrac{T_4 - T_1}{T_3 - T_2} = \dfrac{1}{\kappa}\dfrac{T_1}{T_2}\dfrac{T_4/T_1 - 1}{T_3/T_2 - 1}$

PR's: $\dfrac{T_2}{T_1} = \varphi_1^{\kappa - 1}, \quad \dfrac{T_3}{T_2} = \varphi_2, \quad \dfrac{T_4}{T_3} = \left(\dfrac{\varphi_1}{\varphi_2}\right)^{-(\kappa - 1)}$

$\varphi_1 = \dfrac{V_1}{V_2}, \quad \varphi_2 = \dfrac{V_3}{V_2}$

$$\boxed{\frac{W^-}{Q} = 1 - \frac{\varphi_1^{-(\kappa - 1)}}{\kappa} \cdot \frac{\varphi_2^{\kappa} - 1}{\varphi_2 - 1}}$$

6.3.4 Brayton and Rankine Cycles

A Brayton cycle consists of two isentropes and two isobars. Expressions for W^-/Q_2 and W/Q_1 are found in the same way as for Otto and Diesel cycles. In the heat engine mode a higher compression ratio favours the efficiency at which heat is converted into work.

A Rankine cycle is somewhat related to a Brayton cycle. The isobars in the p-V diagram represent here isobaric charging and discharging. We already encountered this cycle as a special case of the heating/compression cycle in Sect. 6.2.4 with $\Delta\bar{S} = 0$ and $W = \Delta\bar{H}n$. Substitution of ideal-gas PR's gives W^- as a function of p_2/p_1.

A non-expansion cycle comprises isobaric charging, isochoric discharging and isobaric discharging. The work output W^- is found from the contour integral in the p-V diagram.

An incomplete expansion cycle is intermediate between the two previous cycles. By considering an incomplete expansion cycle as a sequence of a Rankine cycle and a non-expansion cycle its work output is readily found.

Brayton cycle

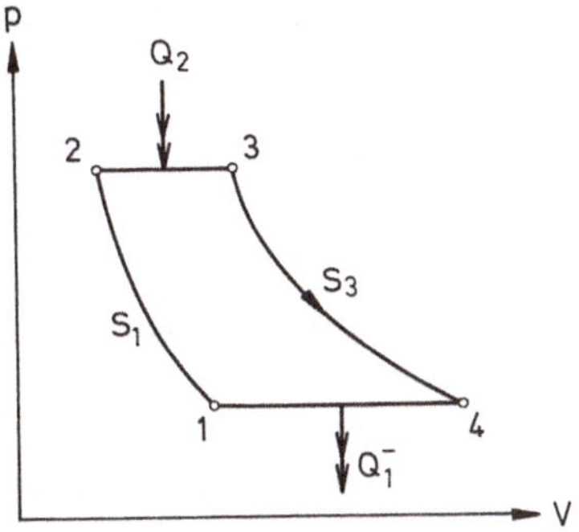

[U]: $0 = Q_2 - Q_1^- - W^-$

$$\frac{W^-}{Q_2} = 1 - \frac{Q_1^-}{Q_2},$$

$$\frac{W}{Q_1} = \frac{Q_2^-}{Q_1} - 1 \text{ (refrigeration mode)}$$

heating/cooling: $\dfrac{Q_1^-}{Q_2} = \dfrac{T_4 - T_1}{T_3 - T_2} = \dfrac{T_1}{T_2}\dfrac{T_4/T_1 - 1}{T_3/T_2 - 1}$

PR's: $\dfrac{T_2}{T_1} = \dfrac{T_3}{T_4} = \left(\dfrac{p_2}{p_1}\right)^{\frac{\kappa-1}{\kappa}}$

$$\boxed{\frac{W^-}{Q_2} = 1 - \frac{T_1}{T_2} = 1 - \frac{T_4}{T_3}}, \quad \boxed{\frac{W}{Q_1} = \frac{T_2}{T_1} - 1 = \frac{T_3}{T_4} - 1}$$

refrigeration mode

Rankine cycle

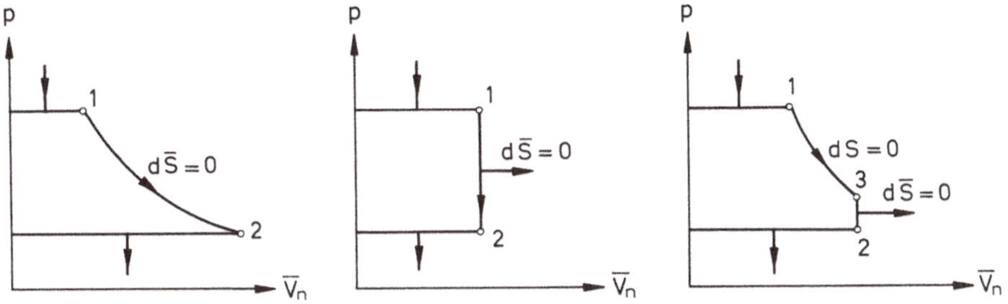

Rankine cycle: $W^- = -\Delta\bar{H}n = n\bar{C}_p(T_1 - T_2) = nRT_1\dfrac{\kappa}{\kappa-1}\left(1 - \dfrac{T_2}{T_1}\right)$

$$= nRT_1\frac{\kappa}{\kappa-1}\left\{1 - \left(\frac{p_2}{p_1}\right)^{\frac{\kappa-1}{\kappa}}\right\}$$

non-expansion cycle: $W^- = \oint p dV = -\oint V dp = n\bar{V}_1(p_1 - p_2)$

$$= nRT_1\left(1 - \frac{p_2}{p_1}\right)$$

incomplete expansion cycle:

$$W^- = nRT_1\left\{\frac{\kappa}{\kappa-1}\left(1 - \frac{T_3}{T_1}\right) + \frac{T_3}{T_1}\left(1 - \frac{p_3}{p_2}\right)\right\}, \quad \frac{T_3}{T_1} = \left(\frac{p_3}{p_1}\right)^{\frac{\kappa-1}{\kappa}}$$

6.3.5 Stirling and Ericsson Cycles

The Stirling and Ericsson cycles conclude the series of reversible ideal-gas cycles.

A Stirling cycle has connected trajects of constant T, V, T and V. In the internal energy balance [U] over a cycle, the isochoric heat input and output cancel each other. Application of the relations for heat rejection/absorption during isothermal compression/expansion yields the expression for W^-/Q. The Stirling cycle can be realised by a cylinder with two pistons with a stationary porous body which exchanges heat with the gas passing through during the isochoric trajects.

The Ericsson cycle can be described in a similar way.

Stirling cycle

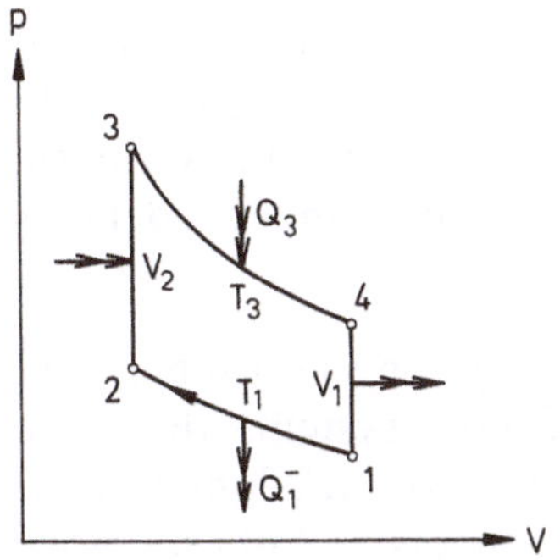

$$[U]:\quad 0 = Q_3 - Q_1^- - W^-$$

$$\frac{W^-}{Q_3} = 1 - \frac{Q_1^-}{Q_3} = 1 - \frac{T_1}{T_3}$$

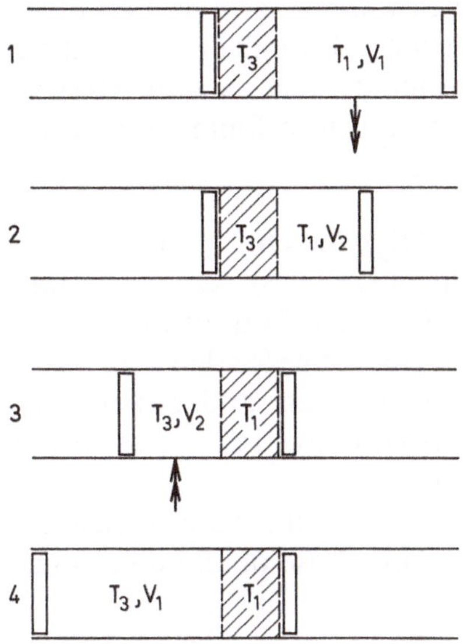

Ericsson cycle

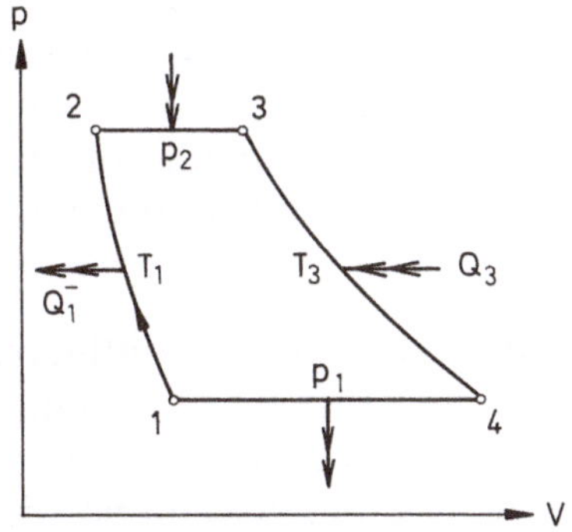

$$[U]:\quad 0 = Q_3 - Q_1^- - W^-$$

$$\frac{W^-}{Q_3} = 1 - \frac{Q_1^-}{Q_3} = 1 - \frac{T_1}{T_3}$$

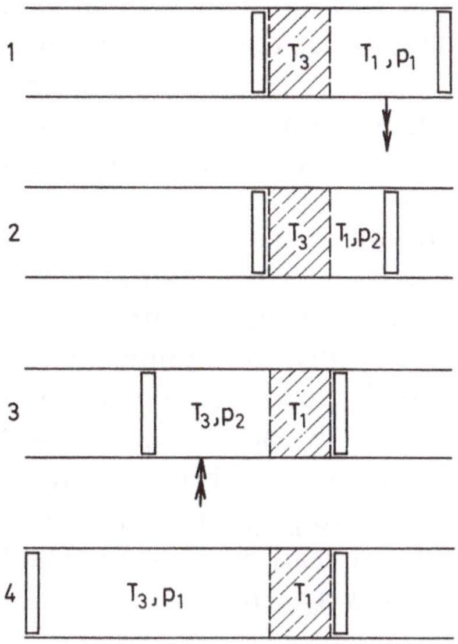

7 BE's of Continuous Plug Flow Systems

We have seen that properties of a non-flow system are time-dependent and place-independent.
In contrast, properties of a continuous flow system are time-independent and place-dependent. This includes the velocities in the flow system. As a consequence all accumulation terms in the describing BE's vanish, while all transport and production rates become time-independent or continuous.

In this chapter we focus on continuous plug flow systems as a subclass of continuous flow systems. In continuous plug flow systems the place dependency of properties is restricted to one direction, the longitudinal direction. At any longitudinal position all components and phases move with the same velocity $\underline{v}$. The properties are gradientless in transversal direction but can exhibit step functions at walls and inlets and outlets.

In plug flow systems with partial or complete recycle the values of the continuous mass, molar and volume flow rates and the time-independent intensive properties pass through a cycle when going in longitudinal direction. Entry and exit values of flow rates and intensive properties are the same for one place-dependent cycle. A once-through system without recycle can be described as a place-dependent cycle as well. Here all transport rates exhibit a step change from zero to a finite value at the entry and return stepwise to zero at the exit.

First we derive a column-2 equation for $d(\delta S_p)$, the entropy production in an infinitesimal stationary element of the plug flow system over a time interval dt using the BE's $[n_i]$, [U] and [S], and a form of the FPR to describe physical equilibrium. The equation has terms for departure from reversible compression, transversal mixing at inlets and outlets, and irreversible chemical reaction. From it an equation for the reversible compression work by external forces emerges.
The FBE's [U] and [S] over a time interval dt and a place-dependent cycle are readily obtained. The FBE's resemble those of an infinitesimal non-flow system.

A time-dependent description of a plug flow system presents itself when following a flow element passing through the system and attaching a time scale to the longitudinal coordinate according to $d\underline{z} = \underline{v}dt$. The description is equivalent to that of an infinitesimal non-flow system or piston-in-cylinder system when the net displacement work on the flow element by the surrounding fluid is included in the work by the piston.

This contribution vanishes when integrating [U] over a time-dependent cycle.

The equivalence of the description as a place-dependent cycle and that of a time-dependent cycle of an infinitesimal non-flow system is illustrated by deriving column-2 equations for Q and W or S_p for heating/cooling and compression/expansion in a continuous plug flow system.

The development of column-2 equations is concluded by applying BE's per unit time to a stationary element of a flow channel for m, m_i, $m\underline{v}$, U + X and S.
Results are the extended Bernoulli equation or mechanical energy balance and the enthalpy balance.
By inserting $d\bar{h}(s, p, w_i)$, a version of the FPR of matter, we obtain an equation for the entropy production rate $d\dot{S}_p$ due to transversal heat transfer and friction at the wall, and irreversible chemical reaction in the flow channel.
By substituting the PR $d\bar{h}(T, p, w_i)$ into the enthalpy balance, the heat balance presents itself. It describes the temperature rise in the flow system due to heat transfer at the wall, wall friction, chemical reaction and reversible compression.

We conclude the chapter on continuous plug flow systems by an example describing a closed continuous Brayton cycle of an ideal gas passing through an isentropic compressor, an isobaric heater, an isentropic turbine and an isobaric heater when in the power generation mode. The equations for the work/heat ratios are the same as for the time-dependent cycles of a non-flow system in Sect. 6.3.4. An open power generation cycle of the Brayton type in a compressor-combustor-turbine combination is described as well.

7.2.1 Description as a Place-Dependent Cycle

In a continuous plug flow system the time-independent properties and continuous rates change with place in one direction, the longitudinal direction. At each place the longitudinal velocity $\underline{v}$ is the same for all components and phases (no longitudinal diffusion and no phase slip). The longitudinal molar and volume flows over a time interval dt, δn and δV, can be expressed in terms of $\underline{v}$ (see Sect. 1.1.5). In a continuous plug flow system with partial or complete recycle the time-independent values of the flows δn and δV and the single-phase or multi-phase properties $\bar{E}$ and $\tilde{E}$ pass through a cycle in the longitudinal direction. Entry and exit values of flows and properties are the same for one place-dependent cycle. A continuous plug flow system without recycle can be described as a place-dependent cycle with $\delta n = 0$ and a stepwise increase of δn at the start of a cycle and a stepwise return to $\delta n = 0$ at the end of a cycle. Transport into and from the system is in transversal direction. The net quantities δn_i, δH, δS, δQ and δW are transported into the system at any one time during a time interval dt at time-independent rates distributed over the place-dependent cycle.

The FBE's of a stationary element of a plug flow system over a time interval dt are readily formulated. The accumulation terms are zero.

In the longitudinal convection term in [U], $-\mathrm{d}(\bar{H}\,\delta n)$, the compression work by the surrounding fluid, the net entry work $-\mathrm{d}(p\delta V)$, is included. The transversal transport terms of the form $\mathrm{d}(\delta E)$ are accompanied by partial molar properties $E_i - \Delta E_i$, where ΔE_i is the property increase in transversal direction when going from the inlet to the element considered. Longitudinal transport of heat is assumed to be zero.

The entropy production in the stationary element is found by combining $[n_i]$, [U] and [S] with the FPR which reflects that physical equilibrium is maintained. Terms for departure from reversible compression, transversal mixing and irreversible reaction can be distinguished.

The reversible external compression work $\mathrm{d}(\mathrm{d}W) = \delta V \mathrm{d}p$ as read from the expression for $T\mathrm{d}(\delta S_p)$ can be interpreted as the work by an external force $\mathrm{d}\underline{F}$ balancing the net pressure force acting on a stationary element.

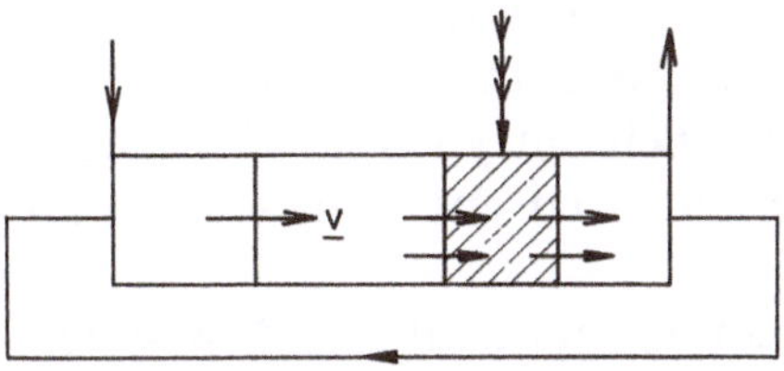

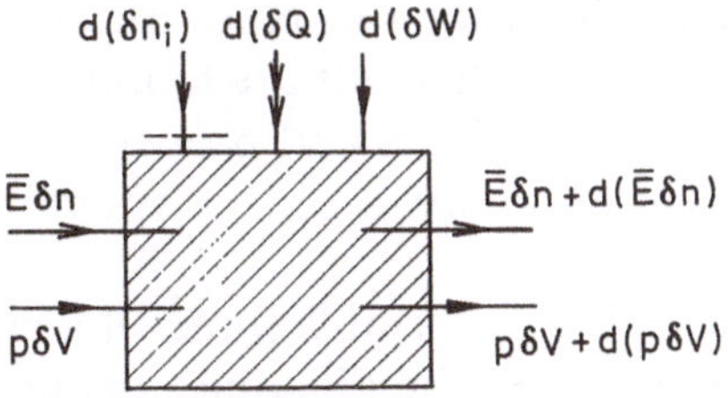

$\delta n = c\underline{v}\,.\,\underline{A}dt$

$\delta V = \underline{v}\,.\,\underline{A}dt = \bar{V}\delta n$

net compression work by fluid: $-\,d(p\delta V)$

FBE's of a stationary element over a time interval dt

$[n_i]$: $0 = -\,d(x_i\delta n) + d(\delta n_i) + \nu_i d(\delta n_{kp})$

$[U]$: $0 = -\,d(\bar{H}\delta n) + d(\delta H) + d(\delta Q) + d(\delta W)$

$$[S]:\quad 0 = -\,d(\bar{S}\delta n) + d(\delta S) + \frac{d(\delta Q)}{T} + d(\delta S_p)$$

FPR: $d\bar{H} = Td\bar{S} + \bar{V}dp + \sum \mu_i dx_i$

Entropy production

$[U] - T[S] - \sum \mu_i [n_i]$:

$$0 = -\,\bar{V}dp\delta n + \sum\{(H_i - \Delta H_i) - T(S_i - \Delta S_i) - \mu_i\}\,d(\delta n_i)$$
$$+\,d(\delta W) - Td(\delta S_p) - \sum \nu_i\mu_i\,d(\delta n_{kp})$$

$$Td(\delta S_p) = \{d(\delta W) - \delta V dp\} + T\sum\left(\Delta S_i - \frac{\Delta H_i}{T}\right)d(\delta n_i) - \sum \nu_i \mu_i\,d(\delta n_{kp})$$

Reversible external compression work

$d(dW) = d\underline{F}\,.\,\underline{v}dt = (\underline{A}dp)\,.\,\underline{v}dt$

$$d(dW) = \delta V dp$$

The FBE's of the place-dependent cycle passing through a string of stationary elements are found by integration over the cycle length recognising that flows and properties return to their entry values after a complete cycle (for a once-through system start and end values of δn are zero).

The FBE's resemble those of a cycle of an infinitesimal non-flow system (Sect. 6.2.2), the difference being that the net quantities δH, δS, δQ and δW, and the entropy production δS_p are here transported and produced over a differential time interval dt and a place-dependent cycle of a finite system rather than a time-dependent cycle of an infinitesimal system. The reversible compression work dW over one cycle equals the contour integral in a p-δV-diagram.

FBE's of a place-dependent cycle

[U]: $$0 = \oint d(\delta H) + \delta Q + \delta W$$

[S]: $$0 = \oint d(\delta S) + \oint \frac{d(\delta Q)}{T} + \delta S_p$$

Reversible compression work

$$-\oint d(p\delta V) = 0$$

$$dW = \oint \delta V dp = -\oint p d(\delta V)$$

7.2.2 Description as a Time-Dependent Cycle

In the previous section we described a continuous plug flow system as a place-dependent cycle. Here we follow an element entering the system at the start of a cycle and flowing through the system at the place-dependent velocity $\underline{v}$. By attaching a time scale to the longitudinal coordinate according to $d\underline{z} = \underline{v}dt$, the description as a place-dependent cycle changes into one as a time-dependent cycle.

We consider a flowing element of length $\underline{v}dt$. In a time interval dt it just displaces its downstream neighbour and receives and rejects the net transversal transport quantities corresponding to $d\underline{z} = \underline{v}dt$. Mole number and volume of the flowing element, dn and dV, are identical to the place-dependent flows over a time interval dt, δn and δV. The longitudinal transport of matter into or from the element travelling with velocity $\underline{v}$ is zero. The sum of the compression work by the surrounding fluid elements $-d(pdV)$ and the external work $d(\delta W)$ can be interpreted as the piston work on a non-flow element.

The FBE's of the flowing element or the substitute non-flow element over a time interval dt are straightforward. The longitudinal convection terms are replaced by accumulation terms, while [U] contains an explicit term for the compression work by the surrounding fluid elements. The equations are equivalent to those of a stationary element as formulated in the previous section when substituting δn and δV for dn and dV, respectively.

By equating the piston work on the non-flow element to the reversible compression work for an infinitesimal non-flow system, $-pd(dV)$, the reversible external compression work is again found.

When we integrate the FBE's over the cycle time t_c in which all properties as well as dn and dV return to their initial values or for which as initial and final values $dn = 0$ and $dV = 0$ apply (plug flow system without recycle), we obtain the description of a continuous plug flow system as a time-dependent cycle. The net compression work by the surrounding fluid elements over a cycle vanishes, while the reversible external compression work dW can be written as a contour integral in a p-dV diagram.

The equivalent description of a continuous plug flow system as a time-dependent cycle of an infinitesimal non-flow element implies that results derived in the previous chapter for non-flow systems can be applied to continuous plug flow systems as well.

Flowing element

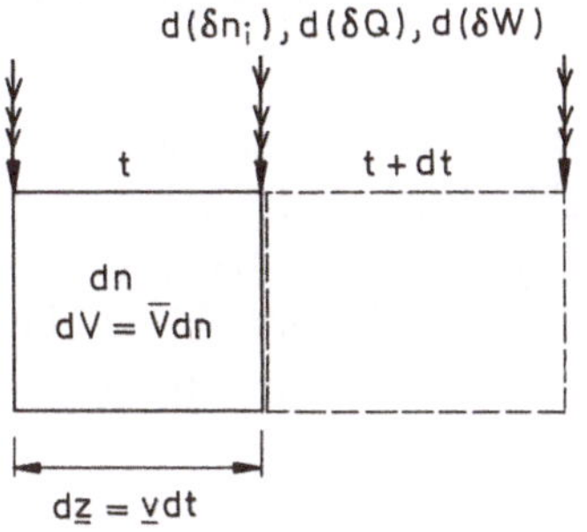

Non-flow element

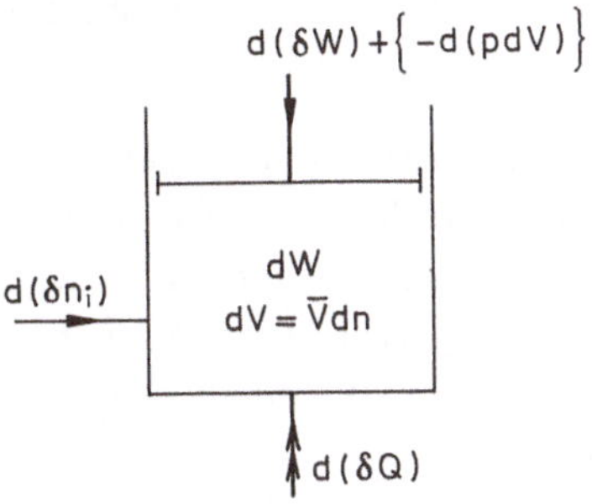

$$dn = c\underline{A}\,.\,\underline{v}dt \equiv \delta n$$

$$dV = \underline{A}\,.\,\underline{v}dt \equiv \delta V$$

net compression work by fluid: $-\,d(p\underline{A}\,.\,\underline{v}dt) = -\,d(pdV) \equiv -\,d(p\delta V)$

FBE's of a flowing element or a non-flow element over a time dt

$[n_i]$: $d(x_i dn) = d(\delta n_i) + \nu_i d(\delta n_{kp})$

$[U]$: $d(\bar{U}dn) = d(\delta H) + d(\delta Q) + d(\delta W) + \{-\,d(pdV)\}$

$[S]$: $d(\bar{S}dn) = d(\delta S) + \dfrac{d(\delta Q)}{T} + d(\delta S_p)$

Reversible external compression work

$$d(dW) + \{-\,d(pdV)\} = -\,pd(dV), \quad d(dW) = dVdp \equiv \delta Vdp$$

FBE's of a time-dependent cycle

$[U]$: $\boxed{0 = \oint d(\delta H) + \delta Q + \delta W}$

$[S]$: $\boxed{0 = \oint d(\delta S) + \oint \dfrac{d(\delta Q)}{T} + \delta S_p}$

Reversible compression work

$$\oint -\,d(pdV) = 0 \qquad \boxed{dW = \oint dVdp = -\oint pd(dV)}$$

7.2.3 Heating/Compression Cycles

We shall now apply the findings of the previous sections to heating or cooling and compression or expansion in a once-through continuous plug flow system.

The p-δV diagram of the place-dependent cycle shows a stepwise increase of the molar flow from zero to δn at the start of the cycle and a stepwise return to zero at the end of the cycle. [U], [S] and the reversible compression work of the cycle are in line with Sect. 7.2.1. The relations for the special case of isothermal compression or expansion follow readily.

As special reversible cases we distinguish continuous isobaric heating/cooling, adiabatic (isentropic) compression/expansion and isothermal compression/expansion.

When comparing the results obtained so far in this section with the findings in Sect. 6.2.4, the analogy between a place-dependent cycle of a continuous plug flow system and a time-dependent cycle of a non-flow system is apparent.

As special irreversible cases we distinguish continuous adiabatic (isenthalpic) expansion with its accompanying Joule–Kelvin effect, and isothermal expansion. Inserting the column-3 PR's of an ideal gas as presented in Sect. 3.3.5, we find the Joule–Kelvin effect to be zero and adiabatic and isothermal expansions to coincide for an ideal gas.

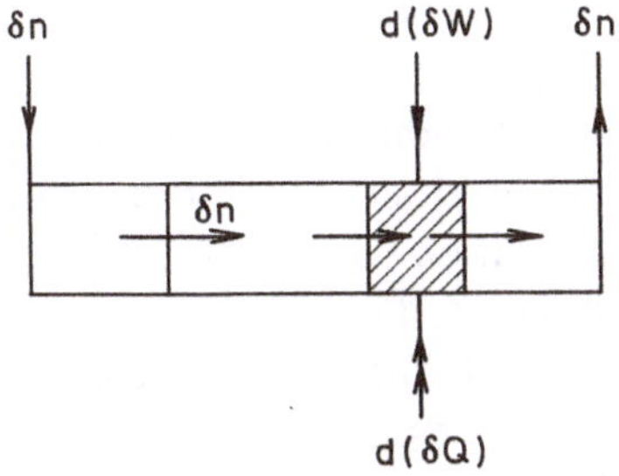

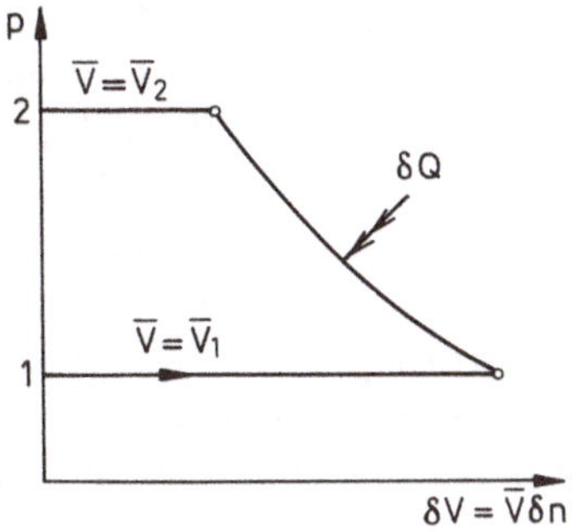

[U]: $0 = -\Delta\bar{H}\delta n + \delta Q + \delta W$

[S]: $0 = -\Delta\bar{S}\delta n + \oint \frac{d(\delta Q)}{T} + \delta S_p$

reversible compression: $dW = \oint \delta V dp = \oint - pd(\delta V)$

Isothermal compression/expansion:

[S]: $0 = -\Delta\bar{S}\delta n + \frac{\delta Q}{T} + \delta S_p$

[U] − T[S]: $0 = -\Delta\bar{G}\delta n + \delta W - T\delta S_p$

Special cases	reversible. $\delta S_p = 0$	irreversible. $\delta W^- = 0$
isobaric heating/ cooling	$dW = 0$ $dQ = \Delta\bar{H}\delta n$	
adiabatic compression/ expansion	$\Delta\bar{S} = 0$ (isentropic) $dW = \Delta\bar{H}\delta n = \int_1^2 \bar{V}dp\,\delta n$	$\Delta\bar{H} = 0$ (isenthalpic) Joule–Kelvin effect: $\left(\frac{\partial T}{\partial p}\right)_{\bar{H}}$ $\delta S_p = \Delta\bar{S}\delta n$
isothermal compression/ expansion	$dW = \Delta\bar{G}\delta n = \int_1^2 \bar{V}dp\,\delta n$ $dQ = T\Delta\bar{S}\delta n$	$T\delta S_p = -\Delta\bar{G}\delta n$ $\delta Q = \Delta\bar{H}\delta n$

7.2.4 BE's per Unit Time

We shall now apply the FBE's per unit time as presented in Sect. 2.2.2 which include the momentum balance $[m\underline{v}]$ and the mechanical energy X in the energy balance $[U + X]$, to a stationary element of a continuous plug flow system. We consider flow in a pipe or duct. Near the wall a film of negligible thickness is assumed in which the longitudinal velocity increases stepwise from zero to a value v. Time-independent transversal transport rates to the element involve transport of longitudinal momentum as a shear force $-\tau_w \, dA_w$ and an external force dF in addition to transport of energy as heat $d\dot{Q}$ and external work $d\dot{W} = dFv$.

First we consider a wall film element with a surface area dA_w. A shear force by the bulk of the fluid on the film balances the shear force by the wall on the film. The work by this force, the friction work $d\dot{W}_{fr}$ appears as a work input in [U] of the wall film element and is rejected as extra heat to the bulk of the fluid. Combination of [U] and [S] yields the rate of entropy production in the wall film element. Terms for the transport of heat from the wall with temperature T_w to the bulk of the fluid with the lower temperature T and for dissipation of work can be distinguished.

Formulation of the FBE's [m], $[m_i]$, [mv], $[U + X]$ and [S] is straightforward. In the momentum balance the longitudinal component of the rate of momentum production, the force by the gravitational field, contains the change in height dh. The two column-2 PR's, a form of the FPR and $d\bar{h}(T, p, w_i)$, are applied below in deriving an expression for the entropy production and in developing the heat balance.

FBE's of wall-film element

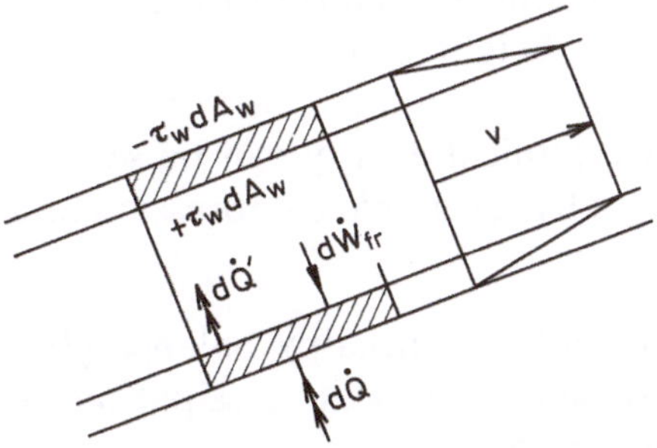

[U]: $0 = d\dot{Q} - d\dot{Q}' + d\dot{W}_{fr}$ $\boxed{d\dot{W}_{fr} = \tau_w v dA_w}$

[S]: $0 = \frac{d\dot{Q}}{T_w} - \frac{d\dot{Q}'}{T} + d\dot{S}_p$ $\boxed{d\dot{S}_p = d\dot{Q}\left(\frac{1}{T} - \frac{1}{T_w}\right) + \frac{d\dot{W}_{fr}}{T}}$

FBE's of element of plug flows system

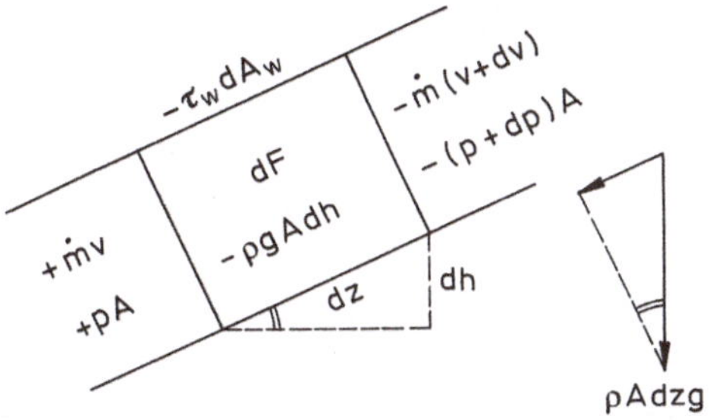

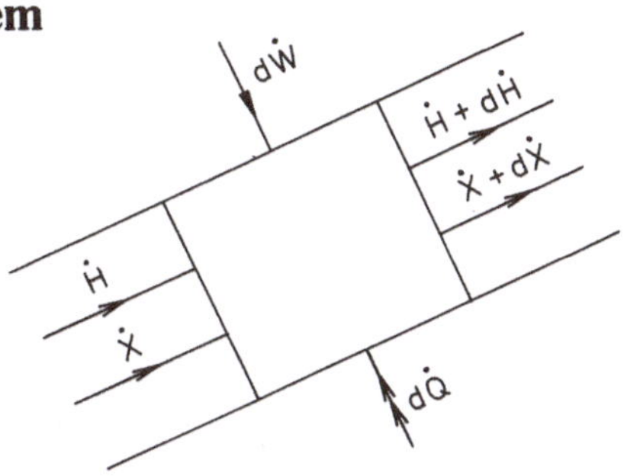

[m]: $0 = -\,d\dot{m} \quad \dot{m} = \rho v A = \text{constant} \quad \dot{V} = vA$

[m_i]: $0 = -\,\dot{m} dw_i + M_i \nu_i d\dot{n}_{kp}$

[mv]: $0 = -\,(\dot{m} dv + A dp + \tau_w dA_w) + dF - \rho g A\, dh$

[U + X]: $0 = -\,(d\dot{H} + d\dot{X}) + d\dot{Q} + d\dot{W}$ $\boxed{d\dot{W} = dFv}$

[S]: $0 = -\,d\dot{S} + \frac{d\dot{Q}}{T_w} + d\dot{S}_p$

PR's: $d\bar{h} = T d\bar{s} + \bar{v} dp + \sum g_i dw_i$

$$d\bar{h} = \bar{c}_p dT + \left(\bar{v} - T\frac{\partial \bar{v}}{\partial T}\right) dp + \sum h_i dw_i$$

The extended Bernoulli equation is obtained as the product of [mv] and v. For zero external work and friction and constant ρ it transforms to the Bernoulli equation which relates velocity, height and pressure changes along a streamline. The extended Bernoulli equation can be rewritten as the mechanical energy balance [X]. Mechanical energy can be seen to increase by external work and to decrease by reversible compression and dissipation of mechanical energy.

By subtracting [X] from [U + X], we find the enthalpy balance [H]. Enthalpy increases by transversal transport of heat, dissipation of mechanical energy and reversible compression.

The rate of entropy production $d\dot{S}_p$ is derived by combining [H], [S], $[m_i]$ and the FPR. It contains the contributions of irreversible heat transport and dissipation of mechanical energy in the wall film as already found above, in addition to the familiar term for irreversible reaction.

Extended Bernoulli equation, mechanical energy balance

[mv] . v:

$$0 = -\dot{m}\left(v\,dv + g\,dh + \frac{1}{\rho}dp\right) + d\dot{W} - d\dot{W}_{fr}$$

ρ = constant, $d\dot{W} = d\dot{W}_{fr} = 0$, Bernoulli equation:

$$\tfrac{1}{2}\rho v^2 + \rho g h + p = \text{constant}$$

[X]:

$$0 = -d\dot{X} + d\dot{W} - \dot{V}dp - d\dot{W}_{fr} \qquad \bar{x} = \tfrac{1}{2}v^2 + gh$$

$$\dot{X} = \dot{m}\bar{x}$$

Enthalpy balance: [U + X] − [X]

[H]:

$$0 = -d\dot{H} + d\dot{Q} + d\dot{W}_{fr} + \dot{V}dp$$

Entropy production

[H]: $0 = -\dot{m}d\bar{h} + d\dot{Q} + d\dot{W}_{fr} + \dot{m}\bar{v}dp$

[S]: $0 = -\dot{m}d\bar{s} + \dfrac{d\dot{Q}}{T_w} + d\dot{S}_p$

$[m_i]$: $0 = -\dot{m}dw_i + M_i\nu_i d\dot{n}_{kp}$

FPR: $d\bar{h} = Td\bar{s} + \bar{v}dp + \sum g_i dw_i$

$[H] - T[S] - \sum g_i[m_i]$:

$$0 = d\dot{W}_{fr} + d\dot{Q}\left(1 - \frac{T}{T_w}\right) - Td\dot{S}_p - \sum \nu_i\mu_i d\dot{n}_{kp}$$

$$d\dot{S}_p = d\dot{Q}\left(\frac{1}{T} - \frac{1}{T_w}\right) + \frac{d\dot{W}_{fr}}{T} - \frac{\sum \nu_i\mu_i d\dot{n}_{kp}}{T}$$

The heat balance emerges when combining [H], [m_i] and the column-2 PR $d\bar{h}(T, p, w_i)$. It describes the increase of T in the longitudinal direction by transversal transport of heat, dissipation of mechanical energy, heat released by a chemical reaction (positive for an exothermic reaction and negative for an endothermic reaction) and reversible compression.

The BE's [m_i], [mv] and [heat] per unit time and per unit volume result from division by dV = Adz. They reflect zero accumulation, longitudinal convective transport, immaterial transport and source terms.

Heat balance

$$[\mathrm{H}]:\quad 0 = -\dot{m}d\bar{h} + d\dot{Q} + d\dot{W}_{fr} + \dot{m}\bar{v}dp$$

$$[\mathrm{m_i}]:\quad 0 = -\dot{m}dw_i + M_i\nu_i d\dot{n}_{kp}$$

$$\text{PR:}\quad d\bar{h} = \bar{c}_p dT + \left(\bar{v} - T\frac{\partial\bar{v}}{\partial T}\right)dp + \sum h_i dw_i$$

$$[\mathrm{H}] - \sum h_i[\mathrm{m_i}]:$$

$$0 = -\dot{m}\left(\bar{c}_p dT - T\frac{\partial\bar{v}}{\partial T}dp\right) + d\dot{Q} + d\dot{W}_{fr} - \sum \nu_i H_i d\dot{n}_{\kappa p}$$

$$[\text{heat}]:\quad \boxed{0 = -\dot{m}\bar{c}_p dT + d\dot{Q} + d\dot{W}_{fr} + \left(-\sum \nu_i H_i\right)d\dot{n}_{kp} + \dot{m}T\frac{\partial\bar{v}}{\partial T}dp}$$

BE's per unit time and per unit volume

$$dV = Adz, \qquad a_w = \frac{dA_w}{dV}$$

$$[\mathrm{m_i}]:\quad \boxed{0 = -\rho v\frac{dw_i}{dz} + M_i\nu_i\dot{n}'''_{kp}}$$

$$[\mathrm{mv}]:\quad \boxed{0 = -\rho v\frac{dv}{dz} - \left(\frac{dp}{dz} + a_w\tau_w\right) + F''' - \rho g\frac{dh}{dz}}$$

$$[\text{heat}]:\quad \boxed{0 = -\rho v\bar{c}_p\frac{dT}{dz} + a_w(\dot{Q}'' + \tau_w v) + \left(-\sum \nu_i H_i\right)\dot{n}'''_{kp} + v\frac{\partial \ln\bar{v}}{\partial \ln T}\frac{dp}{dz}}$$

7.3.1 Brayton Cycles

We now return to a continuous plug flow cycle with negligible mechanical energy effects. We consider a closed continuous Brayton cycle with two isentropes and two isobars in the power generation mode. It can be visualised as a continuous flow through a cycle comprising an isotropic compressor, an isobaric heater, an isentropic turbine and an isobaric cooler. The cycle can be represented in a p-δV diagram. By reversing all flows and transport the cycle changes to its refrigeration mode.

The description as a place-dependent cycle is straightforward. The work/heat ratios for the ideal-gas cycles equal those of the corresponding time-dependent cycles of a non-flow system as derived in Sect. 6.3.4.

Closed power generation cycle

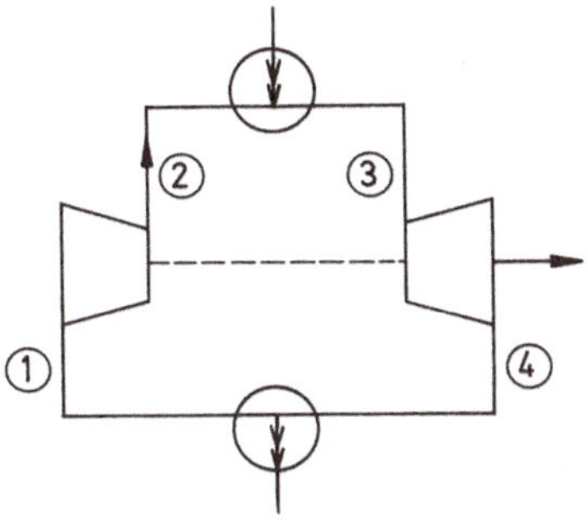

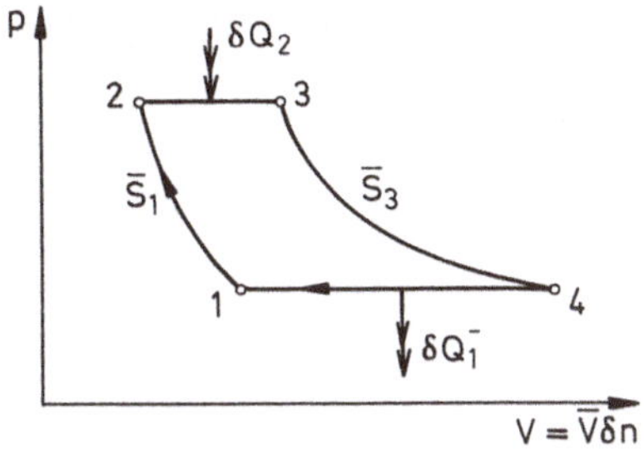

Closed refrigeration cycle

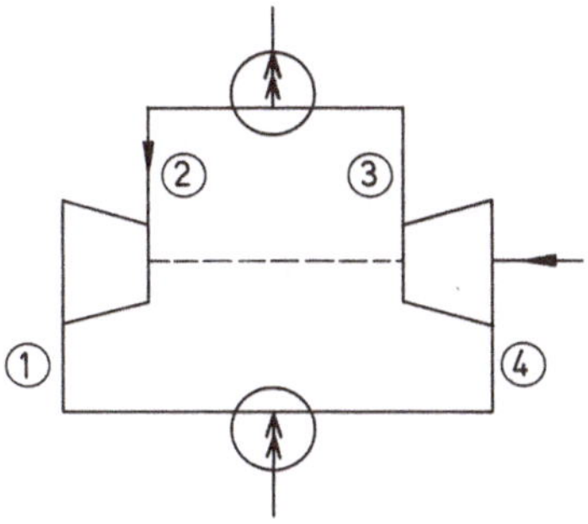

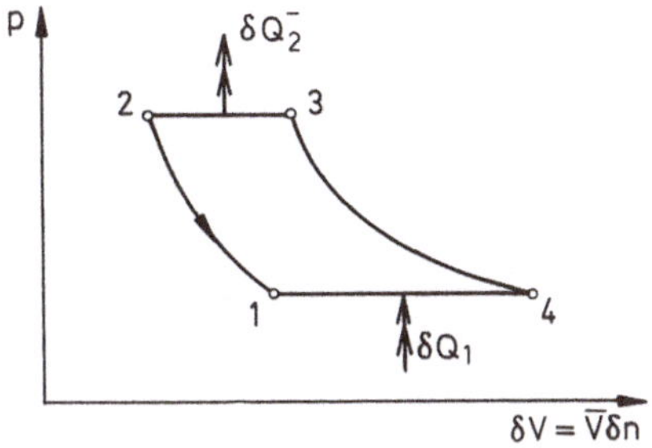

description as place-dependent cycle

$1 \rightarrow 2$ [U]: $0 = -(\bar{H}_2 - \bar{H}_1)\delta n + \delta W_1$

$2 \rightarrow 3$ [U]: $0 = -(\bar{H}_3 - \bar{H}_2)\delta n + \delta Q_2$

$3 \rightarrow 4$ [U]: $0 = -(\bar{H}_4 - \bar{H}_3)\delta n - \delta W_3^-$

$4 \rightarrow 1$ [U]: $0 = -(\bar{H}_1 - \bar{H}_4)\delta n - \delta Q_1^-$

cycle [U]: $0 = (\delta Q_2 - \delta Q_1^-) - (\delta W_3^- - \delta W_1)$

ideal-gas cycles

$$\frac{\delta W^-}{\delta Q_2} = 1 - \frac{T_1}{T_2} = 1 - \frac{T_4}{T_3}, \quad \frac{\delta W}{\delta Q_1} = \frac{T_2}{T_1} - 1 = \frac{T_3}{T_4} - 1$$

An open power generation cycle of the Brayton type can be realised in a compressor-combustor-turbine combination. Its work output equals the enthalpy drop over the system.

Open power generation cycle

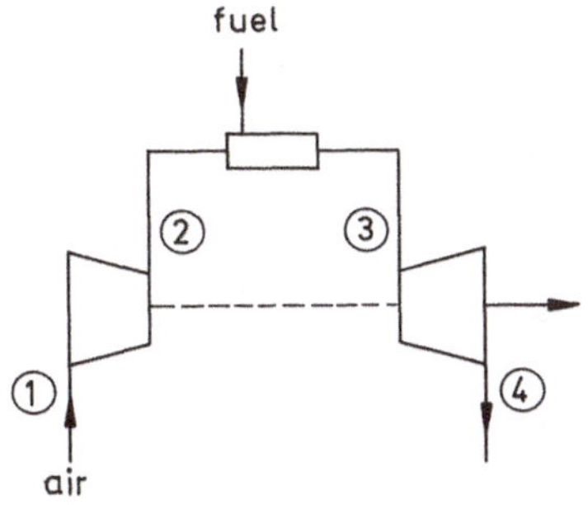

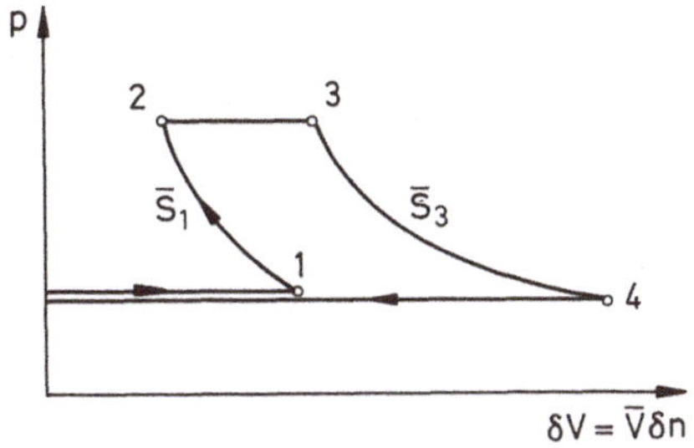

description as place-dependent cycle

isentropic compression: $0 = -(\bar{H}_2 - \bar{H}_1)\delta n + \delta W_1$

isobaric adiabatic combustion: $0 = -\{\bar{H}_3 \delta n' - (\bar{H}_2 \delta n + \bar{H}_f \delta n_f\}$

isentropic expansion: $0 = -(\bar{H}_4 - \bar{H}_3)\delta n' - \delta W_3^-$

cycle: $0 = -\{\bar{H}_4 \delta n' - (\bar{H}_1 \delta n + H_f \delta n_f)\} - \delta W^-$

8 BE's of Continuous Mixed Flow Systems

In Chapter 7 we introduced continuous flow systems having time-independent and place-dependent properties and, as a consequence, zero accumulation terms and continuous transport and production rates in their describing BE's. There we focused on continuous plug flow systems as a subclass. In this chapter we turn to continuous mixed flow systems as a second subclass.

A continuous mixed flow system has place-independent intensive properties. Although properties within the system are gradientless, step changes occur usually at inlets and sometimes at outlets.

The BE's [n_i], [U] and [S] over a time interval yield an enthalpy balance and an equation for δS_p with terms for mixing at the inlet and irreversible chemical reaction.

Next we derive column-2 equations for adiabatic and isothermal mixing/separation. Reversible mixing can be visualised by feeding pure components to a continuous mixed flow system via a cycling infinitesimal piston-in-cylinder expander and a membrane. Directions are reversed to accomplish separation of the mixture. The equations obtained are for work, heat and for entropy production.

Likewise equations are derived for adiabatic and isothermal reaction. A Van't Hoff box with feeding of reactants and withdrawal of products through membranes, presents itself as a device in which reactions can be carried out reversibly.

In Sect. 2.2.6 we described an irreversible reaction in a non-flow system at physical equilibrium. As an exercise we will devise here reversible counterparts which accomplish the same property changes in the non-flow system. They use membranes to withdraw reactants and return products to the non-flow system, piston-in-flow systems for heating/cooling and compression/expansion of reactants and products, Van't Hoff boxes for reversible reaction and Carnot engines and heat pumps to transport heat. The combined work output of the reversibly operating devices equals the exergy loss found before.

The chapter is completed by deriving the heat balance of a continuous backmixed reactor employing as BE's per unit time [m], [m_i], [U] and ideal-gas or ideal-solution PR's.

8.2.1 BE's over a Time Interval

In this section we apply the FBE's over a time interval dt to a continuous mixed flow system. At the entry of the system all intensive properties exhibit a stepwise change to the exit values. The accumulation terms in the FBE's [n_i], [U] and [S] are zero.

The terms of the enthalpy balance [H] can be attributed to mixing at the inlet, chemical reaction and supply of heat to the system.
In the expression for $T\delta S_p$ the by now familiar contributions for irreversible mixing and reaction can be distinguished.

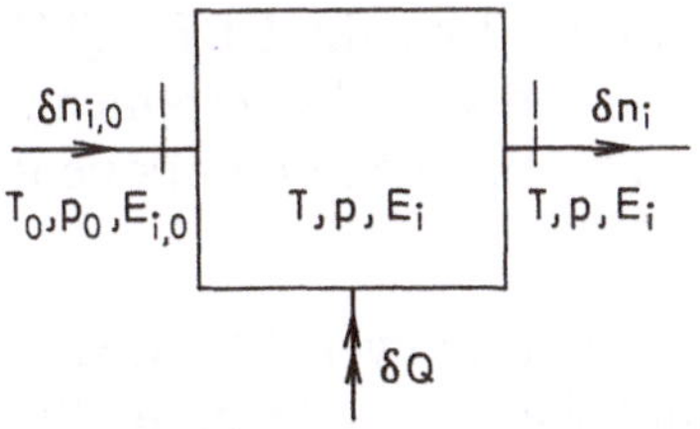

FBE's over a time interval dt

$[n_i]$: $\quad 0 = -(\delta n_i - \delta n_{i,0}) + \nu_i \delta n_{kp}$

$[U]$: $\quad 0 = -\sum\{H_i \delta n_i - (H_i - \Delta H_i)\delta n_{i,0}\} + \delta Q, \quad \Delta E_i = E_i - E_{i,0}$

$[S]$: $\quad 0 = -\sum\{S_i \delta n_i - (S_i - \Delta S_i)\delta n_{i,0}\} + \dfrac{\delta Q}{T} + \delta S_p$

Enthalpy balance

$[U] - \sum H_i [n_i]$:

$$[H]: \quad \boxed{0 = -\sum \Delta H_i \delta n_{i,0} - \sum \nu_i H_i \delta n_{kp} + \delta Q}$$

Entropy production

$[U] - T[S] - \sum \mu_i [n_i]$:

$$0 = -\sum(\Delta H_i - T\Delta S_i)\delta n_{i,0} - T\delta S_p - \sum \nu_i \mu_i \delta n_{kp}$$

$$\boxed{T\delta S_p = T\sum\left(\Delta S_i - \frac{\Delta H_i}{T}\right)\delta n_{i,0} - \sum \nu_i \mu_i \delta n_{kp}}$$

8.2.2 Adiabatic and Isothermal Mixing/Separation

Continuous mixing of pure components is described by the FBE's [U] and [S] over a time interval dt. T and p of the individual components differ in general from T and p of the mixture. The change in component property is denoted by $\Delta E_i = E_i - \bar{E}_i$.

We distinguish reversible mixing with $\delta S_p = 0$ and irreversible mixing with $\delta W^- = 0$. Reversible mixing is accomplished by subjecting the pure-component streams to an expansion cycle in order to obtain the properties for subsequent reversible membrane transport into the mixed system. Reversion of all flows and transports leads to the description of the separation of a mixture into its pure components. Further adiabatic, isothermal and isothermal-isobaric mixing can be distinguished.

Reversible adiabatic mixing is isentropic. Transport through the membranes is isentropic and isenthalpic (see Sect. 2.2.4). The net work output dW^- as found from [U] corresponds with that of an isentropic expansion cycle (Sect. 6.2.4).
Irreversible adiabatic mixing is isenthalpic.

In the case of reversible isothermal mixing, the component free enthalpies are constant when passing the inlet membranes (see Sect. 2.2.4). The net work output dW^- which equals the net influx of G into the system, can be interpreted as the net work output of an isothermal expansion cycle (Sect. 6.2.4). The total heat input dQ for reversible isothermal expansion and mixing is proportional to the net efflux of S from the system.
The expression for $T\delta S_p$ for irreversible isothermal mixing reflects the loss of potential work.

For isothermal-isobaric mixing the property change becomes $\Delta E_{M,i}$. For reversible mixing of ideal gases or liquids $dQ = dW^-$ (zero excess properties).
Irreversible isothermal-isobaric mixing of ideal gases or liquids is adiabatic.

$$[U]: \quad 0 = -\sum \Delta H_i \delta n_i + \delta Q + \delta W \qquad \Delta E_i = E_i - \bar{E}_i$$

$$[S]: \quad 0 = -\sum \Delta S_i \delta n_i + \frac{\delta Q}{T} + \delta S_p$$

	reversible, $\delta S_p = 0$	*irreversible*, $\delta W^- = 0$
	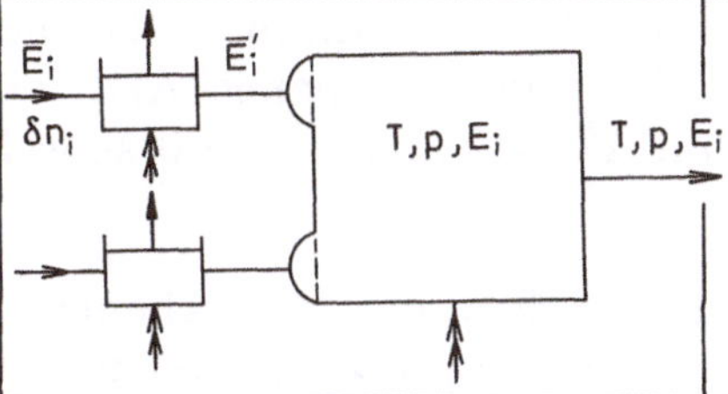	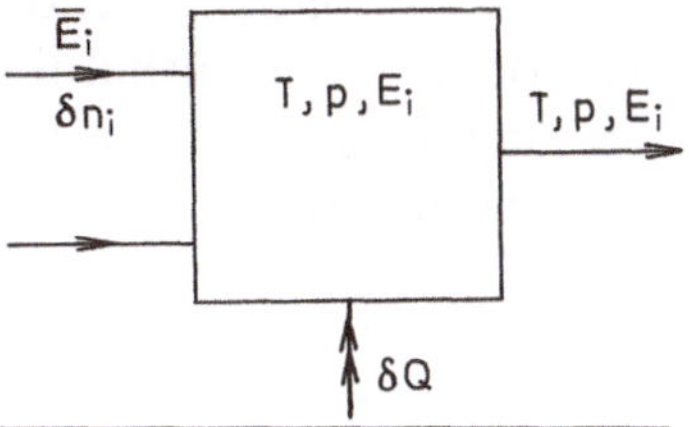
adiabatic	$\bar{S}_i = \bar{S}_i' = S_i, \quad \Delta S_i = 0$ $\bar{H}_i' = H_i, \ \Delta H_i = \bar{H}_i' - \bar{H}_i$ $dW^- = -\sum \Delta H_i \delta n_i$	$\sum \Delta H_i \delta n_i = 0$ $\delta S_p = \sum \Delta S_i \delta n_i$
isothermal	$\bar{G}_i' = \mu_i, \ \Delta G_i = \bar{G}_i' - \bar{G}_i$ $dW^- = -\sum \Delta G_i \delta n_i$ $dQ = T\sum \Delta S_i \delta n_i$	$T\delta S_p = -\sum \Delta G_i \delta n_i$ $\delta Q = \sum \Delta H_i \delta n_i$
isothermal isobaric	$\Delta E_i = E_i(T,p,x_i) - \bar{E}_i(T,p) = \Delta E_{M,i}$ $dW^- = -\sum \Delta G_{M,i} \delta n_i = -(RT\sum x_i \ln x_i + \bar{G}^e)\delta n$ $dQ = T\sum \Delta S_{M,i} \delta n_i = T(-R\sum x_i \ln x_i + \bar{S}^e)\delta n$	$T\delta S_p = -\sum \Delta G_{M,i} \delta n_i = -(RT\sum x_i \ln x_i + \bar{G}^e)\delta n$ $\delta Q = \sum \Delta H_{M,i} \delta n_i = \bar{H}^e \delta n$

8.2.3 Adiabatic and Isothermal Reactions

Here we consider a reaction with one reactant and one product. Reactant and product are continuously supplied to and withdrawn from the mixed flow system as pure compounds by using membranes. Transport through the reactant and product membranes is assumed to be reversible. The pure-component property change by reaction is denoted by $\Delta\bar{E}$, the property change by reaction between the membranes by $\Delta\bar{E}'$.

For an adiabatic reaction the transport through the membranes is isentropic and isenthalpic. The property change $\Delta\bar{H}' = 0$, while for a reversible adiabatic reaction $\Delta\bar{S}' = 0$ and $\Delta\bar{G}' = 0$ as well (adiabatic Van't Hoff box).

For an isothermal reaction the free enthalpy changes across the membranes are zero. When the reaction proceeds reversibly, $\Delta\bar{G}'$ is again zero (isothermal Van't Hoff box).

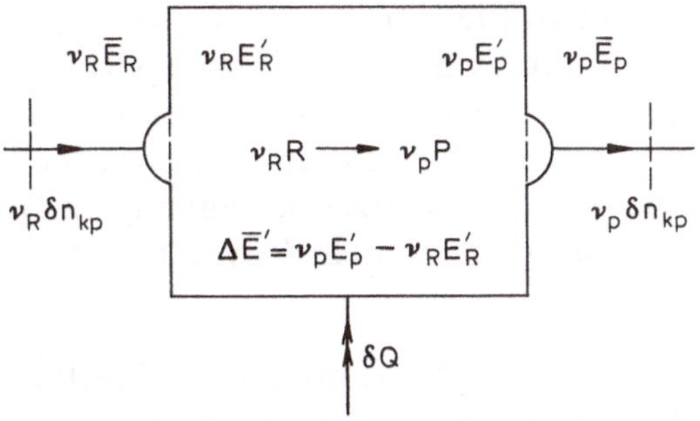

[U]: $0 = -\Delta\bar{H}\delta n_{kp} + \delta Q \qquad \Delta\bar{E} = \nu_p\bar{E}_p - \nu_R\bar{E}_R$

[S]: $0 = -\Delta\bar{S}\delta n_{kp} + \dfrac{\delta Q}{T} + \delta S_p$

	irreversible	*reversible,* $\delta S_p = 0$
adiabatic $\Delta\bar{H} = \Delta\bar{H}'$ $\Delta\bar{S} = \Delta\bar{S}'$	$\Delta\bar{H}' = 0$ $\delta S_p = \Delta\bar{S}'\delta n_{kp}$	$\Delta\bar{H}' = \Delta\bar{S}' = \Delta\bar{G}' = 0$ adiabatic Van't Hoff box
isothermal $\Delta\bar{G} = \Delta\bar{H} - T\Delta\bar{S}$ $= \Delta\bar{G}'$	$\delta Q = \Delta\bar{H}\delta n_{kp}$ $T\delta S_p = -\Delta\bar{G}'\delta n_{kp}$	$dQ = \Delta\bar{H}\delta n_{kp} = T\Delta\bar{S}\delta n_{kp}$ $\Delta\bar{G}' = 0$ isothermal Van't Hoff box

8.2.4 Reversible Counterparts of a Reaction in a Non-Flow System

In Sect. 2.2.6 we derived the expression for the entropy production δS_p in a non-flow system at physical equilibrium with reversible transport into and from the system and between its constituent phases. By equating δS_p to zero we found the condition for chemical equilibrium. Here we shall address the question: "How can the property changes accompanying an irreversible reaction in a non-flow system be matched by a reversible route?"

We consider a reaction with one reactant R and one product P. Reactant with properties H_R and S_R is withdrawn from the non-flow system through a membrane and conditioned in a compression/Carnot heating cycle to enter an adiabatic Van't Hoff box. The product leaving is subjected to an expansion/Carnot cooling cycle to achieve the product properties H_p and S_p prevailing in the non-flow system.

The internal energy balance of the non-flow system relative to that for irreversible reaction, yields an additional heat dQ to balance the net transport of H into the system. The net work output dW^- results from the net exergy influx by matter into the system between the membranes and the Carnot work needed to "pump" the heat quantity dQ into the non-flow system. In total the exergy loss accompanying the irreversible reaction in the non-flow system is here recovered as a work output.

Irreversible reaction

$$\boxed{\delta S_p = -\frac{\Delta\bar{G}}{T}\delta n_{kp}}\,,\quad \Delta\bar{E} = \nu_p E_p - \nu_R E_R$$

Reversible counterpart with an adiabatic Van't Hoff box

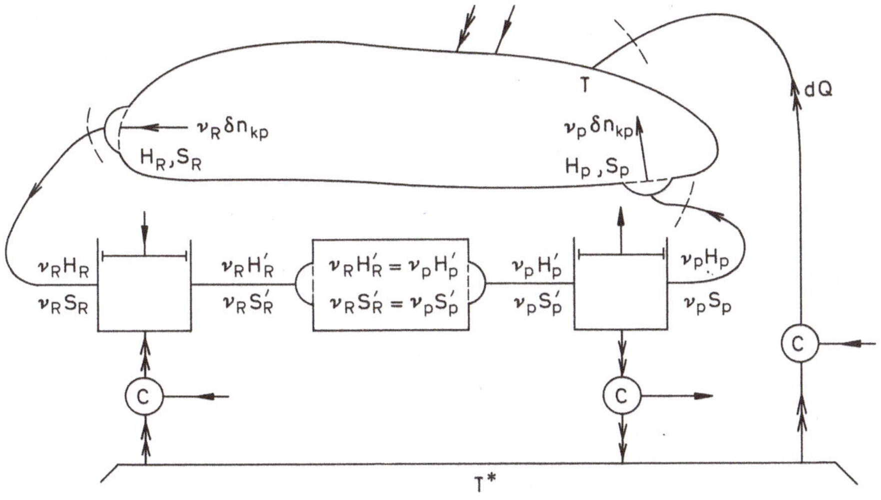

Non-flow system

$[U]_{rev} - [U]_{irr}$: $\quad 0 = \Delta\bar{H}\delta n_{kp} + dQ \qquad \boxed{dQ = -\Delta\bar{H}\delta n_{kp}}$

Net work output

$$dW^- = -(\Delta\bar{H} - T^*\Delta\bar{S})\delta n_{kp} - dQ\left(1 - \frac{T^*}{T}\right)$$

$$= -(\Delta\bar{H} - T^*\Delta\bar{S})\delta n_{kp} + \Delta\bar{H}\left(1 - \frac{T^*}{T}\right)\delta n_{kp}$$

$$= T^*\left(\Delta\bar{S} - \frac{\Delta\bar{H}}{T}\right)\delta n_{kp}$$

$$\boxed{dW^- = -T^*\frac{\Delta\bar{G}}{T}\delta n_{kp}}$$

Another way to recover the exergy loss involves reversible isothermal membrane transport, compression/expansion cycles and reaction in an isothermal Van't Hoff box. The net heat input dQ' needed for the reversible isothermal system equals its net S efflux. The heat dQ into the non-flow system balancing the net U influx follows again from the relative U balance. The net work output dW^- is readily found to be equal to the exergy loss for irreversible reaction.

Reversible counterpart with an isothermal Van't Hoff box

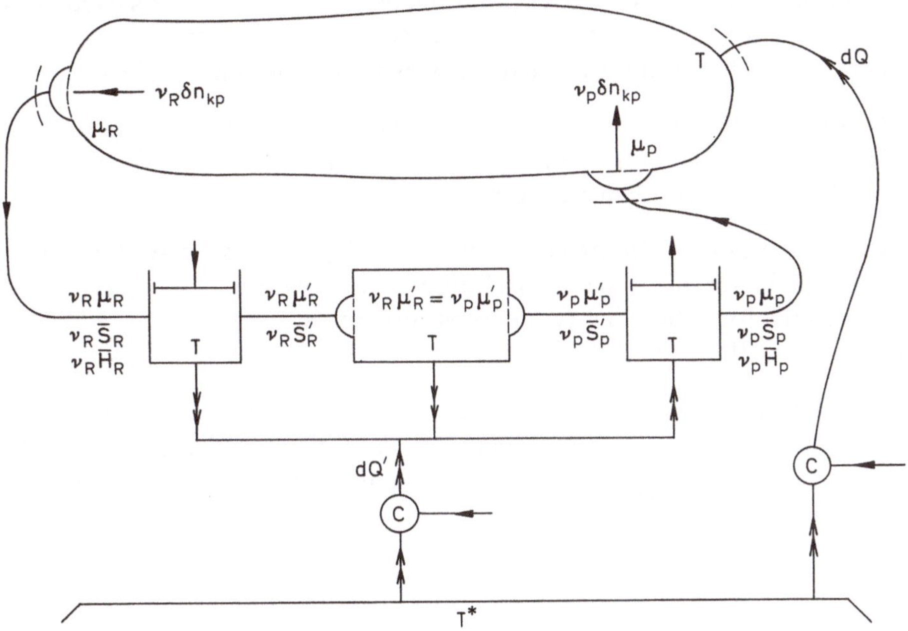

Non-flow system

$[U]_{rev} - [U]_{irr}$:

$$0 = (\nu_p \bar{H}_p - \nu_R \bar{H}_R)\delta n_{kp} + dQ, \quad \boxed{dQ = -(\nu_p \bar{H}_p - \nu_R \bar{H}_R)\delta n_{kp}}$$

Net work output

$$\boxed{dQ' = T(\nu_p \bar{S}_p - \nu_R \bar{S}_R)\delta n_{kp}}$$

$$dW^- = -\Delta\bar{G}\delta n_{kp} - dQ'\left(1 - \frac{T^*}{T}\right) - dQ\left(1 - \frac{T^*}{T}\right)$$

$$= -\Delta\bar{G}\delta n_{kp} + (\nu_p \mu_p - \nu_R \mu_R)\left(1 - \frac{T^*}{T}\right)\delta n_{kp}$$

$$\boxed{dW^- = -T^* \frac{\Delta\bar{G}}{T}\delta n_{kp}}$$

8.3.1 BE's per Unit Time

Here we apply the FBE's per unit time, [m], [m_i] and [U] to a continuous mixed flow system. The continuous heat rate $\dot{Q}$ into the system and the rate of work $\dot{W}'$ by an agitator appear as input terms in [U]. The operation of the continuous mixed flow reactor is assumed to be isobaric.

The BE [m_i] can be rearranged to a form containing the continuous rate of conversion to k per unit mass, $\dot{n}_{kp}/m$ which is a function of T, p and w_i, and the average residence time τ.

Into the column-2 PR at constant p, $d\bar{h}(T, p, w_i)$, the w_i-independent values of h_i for ideal gases or liquids are substituted to obtain two expressions for $\bar{h}$-$\bar{h}_0$. The two expressions reflect two different paths as indicated in the T-w_i diagram.

Accordingly two forms of the heat balance can be formulated. The temperature can be seen to increase by the rate of heat supply $\dot{Q}$ (negative when heat is withdrawn), dissipation of work $\dot{W}'$ and heat of reaction.

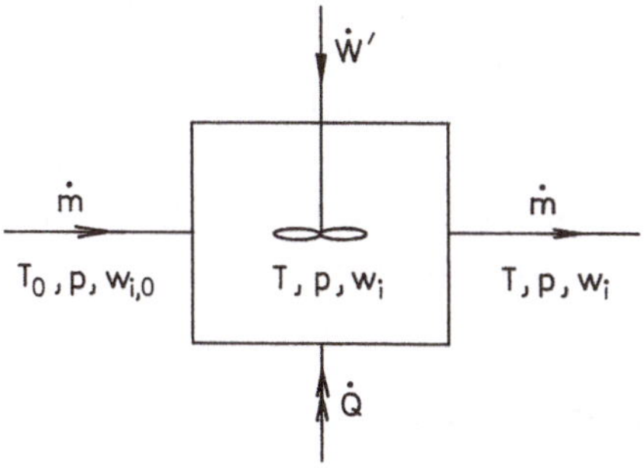

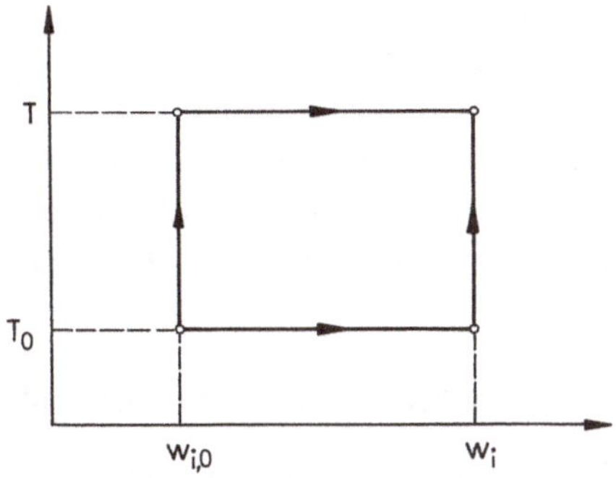

[m]: $0 = -(\dot{m} - \dot{m}_0), \quad \dot{m} = \dot{m}_0$

$[m_i]$: $0 = -\dot{m}(w_i - w_{i,0}) + M_i \nu_i \dot{n}_{kp}$

$$\boxed{0 = -(w_i - w_{i,0}) + M_i \nu_i \frac{\dot{n}_{kp}}{m} \tau} \qquad \tau = \frac{m}{\dot{m}}$$

[U]: $0 = -\dot{m}(\bar{h} - \bar{h}_0) + \dot{Q} + \dot{W}'$

PR: $d\bar{h} = \bar{c}_p dT + \sum h_i dw_i$

ideal gas or ideal solution: $h_i(T,p,w_i) = \bar{h}_i(T,p)$

$$\bar{h} - \bar{h}_0 = \bar{c}_{p,0}(T - T_0) + \sum \bar{h}_i (w_i - w_{i,0})$$
$$= \sum \bar{h}_{i,0}(w_i - w_{i,0}) + \bar{c}_p (T - T_0)$$

Heat balance

$[U] - \sum \bar{h}_i [m_i]$:

$$\boxed{0 = -\dot{m}\bar{c}_{p,0}(T - T_0) + \dot{Q} + \dot{W}' - \sum \nu_i \bar{H}_i \frac{\dot{n}_{kp}}{m} m}$$

$[U] - \sum \bar{h}_{i,0} [m_i]$:

$$\boxed{0 = -\dot{m}\bar{c}_p (T - T_0) + \dot{Q} + \dot{W}' - \sum \nu_i \bar{H}_{i,0} \frac{\dot{n}_{kp}}{m} m}$$

9 BE's of Infinitesimal Systems

In the previous chapters on BE's we focused on non-flow, continuous plug flow and continuous mixed flow systems. These systems have in common that the dependency of properties on time and place is simplified. In contrast we allow properties in this chapter to change both with time and with place in all directions. It is the first chapter in which the linear transport laws postulated in cell 1.3 are applied.

9.1 Column-1 Equations for Infinitesimal Systems

We start to formulate two equivalent BE's for a generalised extensive E to facilitate the formulation of BE's for specific extensive properties.

The BE per unit time and per unit volume of an infinitesimal stationary volume element is referred to as Eulerian. In its transport term the flux of transport of E occurs. As postulated in cell 1.1 transport of E can be decomposed in transport by matter and immaterial transport. The transport by matter is by movement of the individual components with velocity $\underline{v}_i$ relative to the stationary element. It can be further decomposed in convective transport and diffusive transport relative to the mass average velocity $\underline{v}$.
The Eulerian BE for m describes the accumulation of mass in the stationary volume element by net inflow of mass through its boundaries.

The second BE is referred to as Lagrangian. It is the BE per unit time and per unit volume of an infinitesimal mass element following the fluid motion with the mass average velocity $\underline{v}$. The velocity at which the individual components move towards the flowing element is here $\underline{v}_i - \underline{v}$ and accordingly the transport by matter is reduced to diffusive transport only.
The Lagrangian BE for V describes the volume expansion of the flowing mass element due to the net outward displacement of its boundaries.
Using the structure of [E], the BE for m_i is readily formulated.

9.2 Column-2 Equations for Infinitesimal Systems

In this cell we extend first the Lagrangian BE's [E], [V], and [m_i] with the FBE's [$m\underline{v}$], [U + X] and [S]. This is done by utilising the structure of the BE for the generalised extensive and inserting the immaterial transport and the production of fundamental extensives as postulated in cell 1.2.

Two of the FBE's can be rearranged to give the mechanical energy and enthalpy balances.
By multiplying [m$\underline{v}$] with $\underline{v}$ the mechanical energy balance [X] presents itself. It describes the increase in mechanical energy of a flow element following the fluid motion by the net work of pressure and shear forces on its boundaries and by reversible expansion as well as the loss of mechanical energy by dissipation.
[U + X] and [X] are combined to yield the enthalpy balance [H]. This balance describes the increase of H of a flow element by net influx of matter and heat, by dissipation of electrical work and mechanical energy and, finally, by reversible compression by the surrounding fluid.

From [H], [S] and [m_i] and equivalents of the FPR of matter to express local physical equilibrium, we find an equation for the rate of entropy production per unit volume due to molecular transport of momentum, matter and heat,

$$T\dot{S}_p''' = -\underline{\underline{\tau}}:\underline{\nabla}\underline{v} - \sum \underline{\dot{n}}_i'' . \underline{\nabla}A_i - \underline{\dot{Q}}'' . \underline{\nabla}T/T$$

The equation shows that each transport flux is down the gradient of its potential. In cell 1.3 it forms a basis to postulate the linear transport laws. In this cell the equation for $\dot{S}_p'''$ is applied to derive equilibrium concentration distributions for a dilute suspension in the gravitational field and for two isotopes in a centrifugal field by equating $\underline{\nabla}A_i = 0$ and inserting column-3 ideal-state properties of mixing for μ_i.

The heat balance is derived as a last column-2 equation. It is found from [H], [m_i] and the PR $d\bar{h}(T, p, w_i)$. It describes the temperature rise of a flow element following the fluid motion by mixing, net influx of heat, heat release by a chemical reaction, dissipation of electrical work and mechanical energy, and, finally, by reversible compression.

9.3 Column-3 Equations for Infinitesimal Systems

In this cell the linear transport laws postulated in cell 1.3, are brought into play for the first time. They are substituted into the momentum, mole and heat balances to find equations which describe $\underline{v}$, c_i and T as functions of time and place.

The linear law for momentum transport by shear in a newtonian fluid, Newton's law,

$$\underline{\underline{\tau}} = -\mu\{\underline{\nabla}\underline{v} + (\underline{\nabla}\underline{v})^T\},$$

yields upon substitution in the momentum balance [$m\underline{v}$] the Navier–Stokes equation describing $\underline{v}$ as a function of time and place. Under special conditions the momentum balance reduces to the Euler and Bernoulli equations.

The linear law for transport of matter describes multi-component diffusion as a function of the gradients of the chemical potentials μ_i, pressure p, electrical potential ϕ and temperature T. It is first explored for a number of cases.
For a binary mixture we derive a rigorous equation for the flux due to concentration diffusion, pressure diffusion, forced diffusion of ions in an electrical field, and to thermodiffusion.
For components in a binary mixture and dilute components in a multi-component mixture, the linear law for transport of matter reduces for concentration diffusion to Fick's law,

$$\underline{\dot{n}}''_{iD} = - c\mathbb{D}_i\underline{\nabla}x_i.$$

As another example we describe forced diffusion in an ideal solution near an electrode.

When we substitute Fick's law and assume c = constant and $\mathbb{D}_i$ = constant, the mole balance [n_i] for a dilute component turns into a simplified version describing c_i as a function of time and place.

When ignoring the Dufour or diffusion-thermo effect, the linear law for transport of heat reduces to Fourier's law for heat conduction,

$$\underline{\dot{Q}}'' = - \lambda\underline{\nabla}T.$$

A simplified heat balance to describe changes of T is found when using Fourier's law, ignoring diffusion and compression effects and assuming λ = constant.

We conclude the chapter with the set of simplified equations for transport of momentum, matter and heat. They are particularly useful in describing momentum, mass and heat transfer at interfaces and walls. They include the simplified momentum, mole and heat balances with terms for accumulation, convection and molecular transport but without production terms. Further they include the laws of Newton, Fick and Fourier for molecular transport as presented before. In addition they comprise the definition of coefficients for transversal transfer: the Fanning friction coefficient f, the mass transfer coefficient k_i and the heat transfer coefficient α.

The inherent analogy between the equations for momentum, matter and heat is utilised in two examples.

In the first example only accumulation and molecular transport play a role in describing the establishment of a new equilibrium of a system subjected to a step change in $\underline{v}$, c_i or T at its boundaries. Each case can be described by the same partial differential equation in dimensionless form describing the relative departure from equilibrium as a function of dimensionless coordinates and the dimensionless time $\mathscr{D}t/L^2$. The generalised diffusivity $\mathscr{D}$ stands for the kinematic viscosity ν, the diffusivity $\mathbb{D}_i$ or the thermal diffusivity a.

In the second example we consider time-independent transport at an interface or a wall. Without solving any equation a number of conclusions can be drawn.
For geometrically similar cases the dimensionless velocity field as well as the Fanning friction factor f are entirely determined by the dimensionless number Reynolds $Re = v_\infty L/\nu$.
For mass transfer the Sherwood number $Sh = k_i L/\mathbb{D}_i$ is found to be a function of both Re and the Schmidt number $Sc = \nu/\mathbb{D}_i$. Usually the function is of the form $Sh = cRe^m Sc^n$, where c and the exponents are constants.

For heat transfer the Nusselt number $Nu = \alpha L/\lambda$ and the Prandtl number $Pr = \nu/a$ play a role. From the analogy between the equations it follows that Nu is the same function of Re and Pr as Sh is of Re and Sc.

9.1.1 Generalised BE's

In this section we formulate two equivalent BE's of an infinitesimal system for a generalised extensive E. One considers a stationary volume element and is referred to as Eulerian, the other considers a mass element following the fluid motion with the mass average velocity $\underline{v}$ and is called Lagrangian.

We start with a fixed stationary volume element. The accumulation and production terms per unit time and per unit volume are straightforward. The net transport of E per unit time and per unit volume into the element can be expressed as the decrease of the 1-component of the transport flux over a unit length in the 1-direction summed over the three orthogonal coordinate directions (repeated coordinate indices imply summation over the three directions). Alternatively the net influx of E per unit volume can be expressed as the negative of the divergence of vector $\dot{\underline{E}}''_x$, the scalar product of the vector differential operator $\underline{\nabla}$ and $\dot{\underline{E}}''_x$.
We saw in cell 1.1 that transport of E can be decomposed in transport by matter and immaterial transport. Transport by matter is by movement of the individual components with velocity $\underline{v}_i$ and can be further decomposed in convection and diffusion relative to the mass average velocity $\underline{v}$. Substituting the three contributions to $\dot{\underline{E}}''_x$, we find the final form of the Eulerian BE for E.
Upon substitution of m, [E] transforms to the continuity equation [m]. It describes the accumulation of mass in the fixed stationary volume element by net inflow of mass through its boundaries.

We now turn to a given mass element flowing with velocity $\underline{v}$ and formulate the Lagrangian BE for E per unit time and per unit volume. The production term remains unchanged, while the accumulation term becomes the rate of accumulation of E per unit mass in the given mass element multiplied by the prevailing density to obtain the value per unit volume. The individual components move relative to the moving mass element with velocity $\underline{v}_i - \underline{v}$ rather than $\underline{v}_i$. Accordingly transport of E by matter is reduced to diffusive transport relative to the mass average velocity $\underline{v}$.
When comparing the two BE's for E, it is found that for any extensive property accumulation in a moving mass element (without convection through its moving boundaries) equals accumulation in a stationary volume element plus net convective influx. For V this results in [V], which describes the expansion of the mass element due to net outward displacement of its boundaries.
The Lagrangian BE's [E] and [V] are readily extended with [m_i], using the structure of the generalised BE.

Eulerian description

[E]: $$\frac{\partial \check{E}}{\partial t} = -\frac{\partial}{\partial x_l}(\dot{E}''_x)_l + \dot{E}'''_p$$

$$\frac{\partial \check{E}}{\partial t} = -\underline{\nabla}.\underline{\dot{E}}''_x + \dot{E}'''_p$$

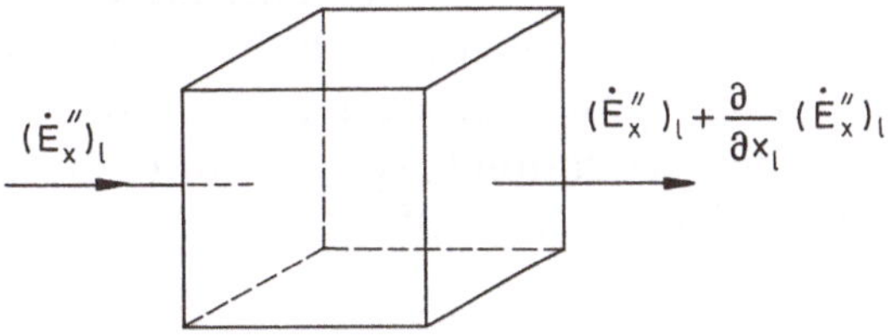

transport by matter:

$$\sum \check{E}_i \underline{v}_i = \check{E}\underline{v} + \sum \check{E}_i(\underline{v}_i - \underline{v}) = \check{E}\underline{v} + \sum e_i \underline{\dot{m}}''_{iD}$$

$$\boxed{\frac{\partial \check{E}}{\partial t} = -\underline{\nabla}.(\check{E}\underline{v} + \sum e_i \dot{m}''_{iD} + \underline{\dot{E}}''_{im}) + \dot{E}'''_p}$$

[m]: $$\boxed{\frac{\partial \rho}{\partial t} = -\underline{\nabla}.\rho\underline{v}}$$

Lagrangian description

[E]: transport by matter: $\sum \check{E}_i(\underline{v}_i - \underline{v}) = \sum e_i \underline{\dot{m}}''_{iD}$

$$\boxed{\rho\frac{D\bar{e}}{Dt} = -\underline{\nabla}.(\sum e_i \underline{\dot{m}}''_{iD} + \underline{\dot{E}}''_{im}) + \dot{E}'''_p}$$

$$\rho\frac{D\bar{e}}{Dt} = \frac{\partial \check{E}}{\partial t} + \underline{\nabla}.\check{E}\underline{v}$$

[V]: $$\boxed{\rho\frac{D\bar{v}}{Dt} = \underline{\nabla}.\underline{v}}$$

[m_i]: $$\boxed{\rho\frac{Dw_i}{Dt} = -\underline{\nabla}.\underline{\dot{m}}''_{iD} + M_i \nu_i \dot{n}'''_{kp}}$$

Above we introduced the substantial derivative $D\bar{e}/Dt$ to denote the rate of accumulation of E per unit mass or the change of $\bar{e}$ with time for a mass element moving with the mass average velocity $\underline{v}$. When using the coupling between Eulerian and Lagrangian accumulation terms and the continuity equation [m], an identity for the substantial derivative of any intensive presents itself.

Substantial derivative

$$\boxed{\rho\frac{D\bar{e}}{Dt}=\frac{\partial\check{E}}{\partial t}+\underline{\nabla}.\check{E}\underline{v}}$$

$$\rho\frac{D\bar{e}}{Dt}=\frac{\partial}{\partial t}(\rho\bar{e})+\underline{\nabla}.(\rho\underline{v}\bar{e})$$

$$=\bar{e}\left(\frac{\partial\rho}{\partial t}+\underline{\nabla}.\rho\underline{v}\right)+\rho\frac{\partial\bar{e}}{\partial t}+\rho\underline{v}.\underline{\nabla}\bar{e}$$

$$=\rho\left(\frac{\partial\bar{e}}{\partial t}+\underline{v}.\underline{\nabla}\bar{e}\right)$$

$$\boxed{\frac{D\bar{e}}{Dt}=\frac{\partial\bar{e}}{\partial t}+\underline{v}.\underline{\nabla}\bar{e}}$$

$$\frac{D\bar{e}}{Dt}=\frac{\partial\bar{e}}{\partial t}+\frac{D\underline{\underline{x}}}{Dt}.\underline{\nabla}\bar{e}=\frac{\partial\bar{e}}{\partial t}+\frac{Dx_l}{Dt}\frac{\partial\bar{e}}{\partial x_l}$$

9.2.1 FBE's

In the previous section we formulated the Lagrangian BE's [E], [V] and [m_i]. The FBE's [$m\underline{v}$], [U + X] and [S] have the same structure as [E] and are obtained by applying immaterial transport fluxes and the volumetric production rates of the fundamental extensives as postulated in cell 1.2.

First we formulate [$m\underline{v}$]. Immaterial transport of momentum is by short-range forces, i.e. pressure and shear forces. The net influx per unit volume is composed of $-\underline{\nabla}p$ and $-\underline{\nabla}.\underline{\underline{\tau}}$, the negatives of the pressure gradient and the divergence of the shear-stress or momentum-flux tensor. The production of momentum is by the long-range force per unit volume due to the gravitational field.

In [U + X] the flux of total transport of energy is composed of diffusive transport of U, the heat flux $\underline{\dot{Q}}''$, the flux of work by pressure forces $p\underline{\dot{V}}''$ and the flux of work by shear forces $\underline{\underline{\tau}}.\underline{v}$. The work by pressure forces can be further split in displacement work and entry work (see Sect. 1.2.2). The transport term in [U + X] is the net influx of the total transport of energy. Production of energy is by work of long-range electrical forces on charged species.

Finally, the entropy balance [S] is readily formulated.

[E]: $$\boxed{\rho \frac{D\bar{e}}{Dt} = -\underline{\nabla}.\left(\sum e_i \underline{\dot{m}}''_{iD} + \underline{\dot{E}}''_{im}\right) + \dot{E}'''_p}$$

$$\rho \frac{D\bar{e}}{Dt} = \frac{\partial \check{E}}{\partial t} + \underline{\nabla}.\check{E}\underline{v}$$

$$\left(\frac{D\bar{e}}{Dt} = \frac{\partial \bar{e}}{\partial t} + \underline{v}.\underline{\nabla}\bar{e}\right)$$

[V]: $$\boxed{\rho \frac{D\bar{v}}{Dt} = \underline{\nabla}.\underline{v}}$$

[m_i]: $$\boxed{\rho \frac{Dw_i}{Dt} = -\underline{\nabla}\cdot\underline{\dot{m}}''_{iD} + M_i \nu_i \dot{n}'''_{kp}}$$

[$m\underline{v}$]: $$\rho \frac{D\underline{v}}{Dt} = -\frac{\partial}{\partial x_l}\left(p\underline{\delta}_l + \tau_{lm}\underline{\delta}_m\right) + \rho \underline{g}$$

$$\boxed{\rho \frac{D\underline{v}}{Dt} = -\left(\underline{\nabla} p + \underline{\nabla}\cdot\underline{\underline{\tau}}\right) + \rho \underline{g}}$$

[U + X]: $$\rho \frac{D}{Dt}(\bar{u} + \bar{x}) = -\underline{\nabla}.\left(\sum u_i \underline{\dot{m}}''_{iD} + \underline{\dot{Q}}'' + p\underline{\dot{V}}'' + \underline{\underline{\tau}}.\underline{v}\right) + \sum \underline{\dot{m}}''_{iD}.\underline{f}_i$$

$$p\underline{\dot{V}}'' = p\underline{v} + \sum p v_i \underline{\dot{m}}''_{iD}$$

$$\boxed{\rho \frac{D}{Dt}(\bar{u} + \bar{x}) = -\underline{\nabla}.\left(\sum h_i \underline{\dot{m}}''_{iD} + \underline{\dot{Q}}'' + p\underline{v} + \underline{\underline{\tau}}.\underline{v}\right) + \sum \underline{\dot{m}}''_{iD}.\underline{f}_i}$$

[S]: $$\boxed{\rho \frac{D\bar{s}}{Dt} = -\underline{\nabla}.\left(\sum s_i \underline{\dot{m}}''_{iD} + \frac{\underline{\dot{Q}}''}{T}\right) + \dot{S}'''_p}$$

9.2.2 Mechanical Energy and Enthalpy Balances

In this section we shall derive the BE's [X] and [H] starting from [m$\underline{v}$] and [U + X] as presented in the previous section.

The scalar product of [m$\underline{v}$] and $\underline{v}$ yields a BE for the kinetic energy. The kinetic energy can be seen to increase as a result of work by the resultant of the short-range forces acting on the boundaries of an element and as a result of work by the gravitational field. The latter can be rewritten as the decrease of potential energy in the gravitational field. This gives rise to the first form of [X] which shows the total mechanical energy of an element with velocity $\underline{v}$ to increase as a result of work by the resultant of the short-range forces acting on its boundaries.

The final form of [X] is obtained by splitting the work by the net pressure force and the work by the net shear force on an element in two parts. The mechanical energy of an element following the fluid motion increases by the net displacement and shear work by the surrounding fluid and by reversible expansion. The scalar product $\underline{\underline{\tau}} : \underline{\nabla}\underline{v}$ will be found to be always negative. It represents the irreversible conversion (dissipation) of mechanical energy to internal energy.

When we combine the energy balance [U + X] with the mechanical energy balance [X], the enthalpy balance [H] emerges. The Lagrangian accumulation of enthalpy is by net influx of matter and heat, dissipation of electrical work and mechanical energy, and reversible compression.

Mechanical energy balance

[V]: $$\rho\frac{D\bar{v}}{Dt} = \underline{\nabla}\cdot\underline{v}$$

[m$\underline{v}$]: $$\rho\frac{D\underline{v}}{Dt} = -(\underline{\nabla}p + \underline{\nabla}\cdot\underline{\underline{\tau}}) + \rho\underline{g}$$

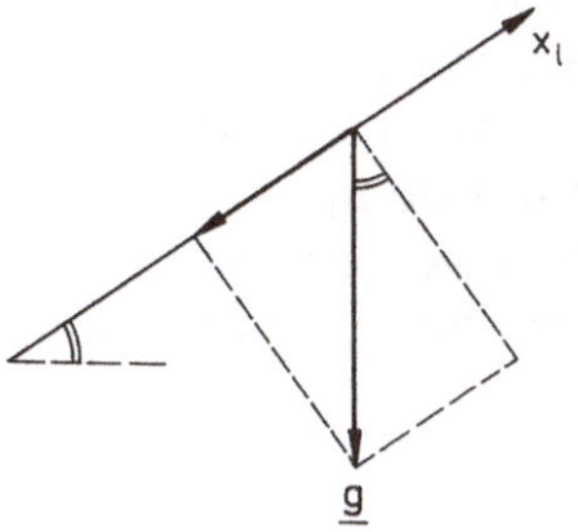

[m$\underline{v}$]$\cdot\underline{v}$: $$\rho\frac{D}{Dt}(\tfrac{1}{2}v^2) = -(\underline{\nabla}p + \underline{\nabla}\cdot\underline{\underline{\tau}})\cdot\underline{v} + \rho\underline{g}\cdot\underline{v}$$

$$\underline{g} = -g\frac{\partial h}{\partial x_l}\underline{\delta}_l = -\underline{\nabla}(gh)$$

$$\rho\underline{g}\cdot\underline{v} = -\rho\underline{v}\cdot\underline{\nabla}(gh)$$

$$= -\rho\left\{\frac{\partial}{\partial t}(gh) + \underline{v}\cdot\underline{\nabla}(gh)\right\} = -\rho\frac{D}{Dt}(gh)$$

[X]: $$\boxed{\rho\frac{D\bar{x}}{Dt} = -(\underline{\nabla}p + \underline{\nabla}\cdot\underline{\underline{\tau}})\cdot\underline{v}}$$

$$-\underline{\nabla}p\cdot\underline{v} = -\left(\frac{\partial}{\partial x_l}p\right)v_l = -\frac{\partial}{\partial x_l}(pv_l) + p\frac{\partial}{\partial x_l}v_l = -\underline{\nabla}\cdot(p\underline{v}) + p\underline{\nabla}\cdot\underline{v}$$

$$= -\underline{\nabla}\cdot(p\underline{v}) + \rho p\frac{D\bar{v}}{Dt}$$

$$-(\underline{\nabla}\cdot\underline{\underline{\tau}})\cdot\underline{v} = -\left(\frac{\partial}{\partial x_l}\tau_{lm}\right)v_m = -\frac{\partial}{\partial x_l}(\tau_{lm}v_m) + \tau_{ml}\frac{\partial}{\partial x_l}v_m$$

$$= -\underline{\nabla}\cdot(\underline{\underline{\tau}}\cdot\underline{v}) + \underline{\underline{\tau}}:\underline{\nabla}\underline{v}$$

[X]: $$\boxed{\rho\frac{D\bar{x}}{Dt} = -\underline{\nabla}\cdot(p\underline{v} + \underline{\underline{\tau}}\cdot\underline{v}) + \rho p\frac{D\bar{v}}{Dt} + \underline{\underline{\tau}}:\underline{\nabla}\underline{v}}$$

Enthalpy balance

[U + X]: $$\rho\frac{D}{Dt}(\bar{u} + \bar{x}) = -\underline{\nabla}\cdot\left(\sum h_i\underline{\dot{m}}''_{iD} + \underline{\dot{Q}}'' + p\underline{v} + \underline{\underline{\tau}}\cdot\underline{v}\right) + \Sigma\underline{\dot{m}}''_{iD}\cdot\underline{f}_i$$

$$\bar{h} = \bar{u} + p\bar{v}$$

[H]: $$\boxed{\rho\frac{D\bar{h}}{Dt} = -\underline{\nabla}\cdot\left(\sum h_i\underline{\dot{m}}''_{iD} + \underline{\dot{Q}}''\right) + \sum\underline{\dot{m}}''_{iD}\cdot\underline{f}_i - \underline{\underline{\tau}}:\underline{\nabla}\underline{v} + \frac{Dp}{Dt}}$$

9.2.3 Entropy Production

The volumetric rate of entropy production follows from a linear combination of [H], [S] and [m_i] when inserting a form of the FPR of matter, $d\bar{h}(\bar{s}, p, w_i)$, to reflect that locally physical equilibrium is maintained. In this expression for $T\dot{S}_p'''$ contributions can be distinguished due to dissipation of mechanical energy, mixing, dissipation of electrical work, heat transport, and an irreversible chemical reaction. Each transport flux occurs in a product with its accompanying potential gradient, the reaction rate $\dot{n}_{kp}'''$ in a product with its potential difference $\Sigma \nu_i \mu_i$. As $T\dot{S}_p'''$ is positive, each flux or rate and its potential gradient or potential difference have opposite signs.

$$[H]: \quad \rho\frac{D\bar{h}}{Dt} = -\underline{\nabla}.\left(\sum h_i \underline{\dot{m}}''_{iD} + \underline{\dot{Q}}''\right) + \sum \underline{\dot{m}}''_{iD}.\underline{f}_i - \underline{\underline{\tau}}:\underline{\nabla}\underline{v} + \frac{Dp}{Dt}$$

$$[S]: \quad \rho\frac{D\bar{s}}{Dt} = -\underline{\nabla}.\left(\sum s_i \underline{\dot{m}}''_{iD} + \frac{\underline{\dot{Q}}''}{T}\right) + \dot{S}'''_p$$

$$[m_i]: \quad \rho\frac{Dw_i}{Dt} = -\underline{\nabla}.\underline{\dot{m}}''_{iD} + M_i \nu_i \dot{n}'''_{kp}$$

$$\text{FPR:} \quad d\bar{h} = Td\bar{s} + \bar{v}dp + \sum g_i dw_i$$

$$dh_i = Tds_i + v_i dp + (dg_i)_{T,p}$$

$[H] - T[S] - \sum g_i [m_i]$:

$$\frac{Dp}{Dt} = -\sum(h_i - Ts_i - g_i)\underline{\nabla}.\underline{\dot{m}}''_{iD} - \sum \underline{\dot{m}}''_{iD}.\{(\underline{\nabla}h_i - T\underline{\nabla}s_i) - \underline{f}_i\}$$

$$-\underline{\nabla}.\underline{\dot{Q}}'' + T\left(\frac{1}{T}\underline{\nabla}.\underline{\dot{Q}}'' - \frac{\underline{\dot{Q}}''}{T^2}.\underline{\nabla}T\right)$$

$$-\underline{\underline{\tau}}:\underline{\nabla}\underline{v} + \frac{Dp}{Dt} - T\dot{S}'''_p - \sum \nu_i \mu_i \dot{n}'''_{kp}$$

$$\boxed{T\dot{S}'''_p = -\underline{\underline{\tau}}:\underline{\nabla}\underline{v} - \sum \underline{\dot{m}}''_{iD}\cdot\{(\underline{\nabla}h_i - T\underline{\nabla}s_i) - \underline{f}_i\} - \underline{\dot{Q}}''.\frac{\underline{\nabla}T}{T} - \sum \nu_i \mu_i \dot{n}'''_{kp}}$$

The entropy production due to diffusion can be further specified by inserting the column-2 PR $dh_i(s_i, p, w_i)$.
As $\Sigma \underline{\dot{m}}''_{iD} = 0$, the entropy production is invariant to the inclusion of a constant $\underline{\nabla} b$ in the gradient of the diffusion potential for each component.
Rewriting the entropy production on a mole basis yields the gradients $\underline{\nabla} A_i$ of the individual components which still contain the unknown constant $\underline{\nabla} b$. The electrical force on an ion is rewritten in terms of its charge z_i and the electrical potential ϕ.
In deriving the entropy production due to diffusion, we considered transport fluxes for individual components relative to a point with the mass average velocity $\underline{v}$. If we postulate the total entropy production due to differences between the component velocities $\underline{v}_i$ to be invariant to the choice of the reference velocity, we find that the mole average of the gradients $\underline{\nabla} A_i$ must be zero. From this condition we can solve the unknown constant $\underline{\nabla} b$ using the Gibbs–Duhem relation for G and the electroneutrality condition for a fluid with electrically charged species.
The local diffusion potential A_i for a component is high where its chemical potential μ_i and, hence, its concentration is high, where p is high when it has a larger than average specific volume, and where the electrical potential ϕ is high when it has a positive charge.

Substituting the modified fluxes and gradients for the transport of matter, we find the final form of the expression for $T\dot{S}'''_p$ for transport of momentum, matter and heat.

$T\dot{S}'''_p = 0$ at equilibrium or when the transport processes proceed reversibly. All gradients are then zero: $\underline{\nabla}\underline{v} = 0$, $\underline{\nabla} A_i = 0$ for each component, and $\underline{\nabla} T = 0$.

In the absence of equilibrium $T\dot{S}'''_p$ is always positive. It follows that each transport flux is down the gradient of its potential: transversal transport of longitudinal momentum is in the direction of lower v_l, transport of n_i is to a lower diffusion potential A_i and transport of heat to a lower T. Starting from the expression for $T\dot{S}'''_p$ experimental linear transport laws for momentum, matter and heat are postulated as column-3 principles in cell 1.3.

Entropy production by transport of matter

$$(T\dot{S}_p''')_D = -\sum \underline{\dot{m}}_{iD}''.\{(\underline{\nabla}g_i)_{T,p} + v_i\underline{\nabla}p - \underline{f}_i\}$$
$$= -\sum \underline{\dot{m}}_{iD}''.\{(\underline{\nabla}g_i)_{T,p} + v_i\underline{\nabla}p - \underline{f}_i + \underline{\nabla}b\} \quad (\sum \underline{\dot{m}}_{iD}'' = 0)$$
$$= -\sum w_i\rho(\underline{v}_i - \underline{v}).\{(\underline{\nabla}g_i)_{T,p} + v_i\underline{\nabla}p - \underline{f}_i + \underline{\nabla}b\}$$
$$= -\sum x_i c(\underline{v}_i - \underline{v}).\{(\underline{\nabla}\mu_i)_{T,p} + V_i\underline{\nabla}p - \underline{F}_i + M_i\underline{\nabla}b\}$$
$$= -\sum x_i c(\underline{v}_i - \underline{v}).\underline{\nabla}A_i$$

$$\underline{\underline{\nabla}A_i = (\underline{\nabla}\mu_i)_{T,p} + V_i\underline{\nabla}p + z_i\mathscr{F}\underline{\nabla}\phi + M_i\underline{\nabla}b}$$

invariance w.r.t. reference velocity:

$$\boxed{\sum x_i\underline{\nabla}A_i = 0}$$

$$\sum x_i\{\underline{\nabla}\mu_i)_{T,p} + V_i\underline{\nabla}p + z_i\mathscr{F}\underline{\nabla}\phi + M_i\underline{\nabla}b\} = 0$$
$$\bar{V}\underline{\nabla}p + \bar{M}\underline{\nabla}b = 0, \quad \underline{\nabla}b = -\bar{v}\underline{\nabla}p$$

$$\boxed{\underline{\nabla}A_i = (\underline{\nabla}\mu_i)_{T,p} + M_i(v_i - \bar{v})\underline{\nabla}p + z_i\mathscr{F}\underline{\nabla}\phi}$$

$$\boxed{(T\dot{S}_p''')_D = -\sum \underline{\dot{n}}_i''.\underline{\nabla}A_i}, \quad \boxed{\underline{\dot{n}}_i'' = x_i c(\underline{v}_i - \underline{v}_{ref})}$$

Entropy production by transport of momentum, matter and heat

$$\boxed{T\dot{S}_p''' = -\underline{\underline{\tau}}:\underline{\nabla}\underline{v} - \sum \underline{\dot{n}}_i''.\underline{\nabla}A_i - \underline{\dot{Q}}''.\frac{\underline{\nabla}T}{T}}$$

$$T\dot{S}_p''' = -\tau_{nl}\frac{\partial}{\partial x_n}v_l - \sum_i \dot{n}_{i,n}''\frac{\partial}{\partial x_n}A_i - \frac{1}{T}\dot{Q}_n''\frac{\partial T}{\partial x_n}$$

9.2.4 Equilibrium Concentration Distributions

Before we bring the column-3 transport laws on the scene, we shall apply the equilibrium conditions $\underline{\nabla} A_i = 0$ to two examples with an equilibrium concentration distribution accompanying a pressure gradient.

The first example refers to a dilute suspension of solid particles in a liquid in the gravitational field. The particles are regarded as molecules and the suspension as a single phase. In the equilibrium condition $\underline{\nabla} A_i = 0$ we substitute the infinite-dilution law (Sect. 1.3.1) for the chemical potential μ_i of the particles and the pressure gradient due to the gravitational field. When the particles are denser than the liquid or $\bar{\rho}_1 > \bar{\rho}_2$, their concentration declines exponentially with the height h. The exponent determining the equilibrium distribution suggests a competition between the potential energy of a particle in the gravity field and the energy kT of a molecule associated with random movement.

The second example concerns the equilibrium distribution of a binary gas mixture in a centrifugal field. In the equilibrium condition $\underline{\nabla} A_i = 0$ we assume for μ_i ideal-mixture behaviour and $V_1 = V_2 = \bar{V}$ which are good approximations when separating e.g. $U^{235}F_6$ and $U^{238}F_6$ in an ultracentrifuge. For a constant angular velocity ω the pressure increases with the radius r due to the centrifugal acceleration $\omega^2 r$. The radial distribution of the mole ratio indicates that the mixture contains less of the lighter isotope at increasing radius. The exponent suggests a competition between the difference in kinetic energy for the two molecules and the energy associated with random molecular movement.

Dilute suspension in gravitational field

$$\underline{\nabla} A_1 = (\underline{\nabla}\mu_1)_{T,p} + M_1(v_1 - \bar{v})\underline{\nabla} p = 0$$

$$RT\,d(\ln x_1) + M_1(\bar{v} - v_1)\rho g\,dh = 0$$

$$\boxed{\ln\frac{x_1(h)}{x_1(0)} \approx -\left(1 - \frac{\bar{\rho}_2}{\bar{\rho}_1}\right)\frac{M_1 gh}{RT} = -\left(1 - \frac{\bar{\rho}_2}{\bar{\rho}_1}\right)\frac{dm_1 gh}{kT}}$$

dm_1 = mass of a particle

$$k = \frac{R}{N_{av}} \qquad \text{(Boltzmann's constant)}$$

$$N_{av} \approx 6 \times 10^{26} \text{ molecules (kmole)}^{-1}$$

Ideal mixture with $V_1 = V_2 = \bar{V}$ in a centrifugal field

$$\underline{\nabla} A_1 = (\underline{\nabla}\mu_1)_{T,p} + M_1(v_1 - \bar{v})\underline{\nabla} p = 0$$

$$RT\,d(\ln x_1) + M_1\left(\frac{\bar{V}}{M_1} - \bar{v}\right)\rho\omega^2 r\,dr = 0$$

$$RT\,d(\ln x_1) + (\bar{M} - M_1)\omega^2 r\,dr = 0$$

$$RT\frac{1}{x_1}dx_1 + x_2(M_2 - M_1)\omega^2 r\,dr = 0$$

$$\frac{1}{x_1 x_2}dx_1 = \left(\frac{1}{x_1} + \frac{1}{x_2}\right)dx_1 = -\frac{(M_2 - M_1)\omega^2 r\,dr}{RT}$$

$$\boxed{\ln\frac{x_1/x_2(r)}{x_1/x_2(0)} = -\frac{(M_2 - M_1)(\omega r)^2}{2RT}}$$

9.2.5 Heat Balance

Here we shall complete the column-2 part of this chapter by deriving the heat balance. The derivation is along the same lines as for a non-flow system (Sect. 6.2.6), a continuous plug flow system (Sect. 7.2.4) and a continuous mixed flow system (Sect 8.3.1). In Sect 6.2.6 we noted already that the term heat balance is not entirely correct as it suggests that heat is an extensive system property rather than a form of energy being transported.

The heat balance is found by combining [H], [m_i] and the PR $d\bar{h}(T, p, w_i)$. The temperature of an element following the fluid motion increases by mixing, net influx of heat, heat release by a chemical reaction, dissipation of electrical work, dissipation of mechanical energy and reversible compression by the surrounding fluid. The contributions involving the diffusion flux $\underline{\dot{m}}''_{iD}$, mixing and dissipation of electrical work, are usually negligible. The effect of reversible compression is to a good approximation negligible for a liquid and zero when p is independent of time and place.

$$[H]: \quad \rho\frac{D\bar{h}}{Dt} = -\underline{\nabla}\cdot\left(\sum h_i \underline{\dot{m}}''_{iD} + \underline{\dot{Q}}''\right) + \sum \underline{\dot{m}}''_{iD}\cdot\underline{f}_i - \underline{\underline{\tau}}:\underline{\nabla}\underline{v} + \frac{Dp}{Dt}$$

$$[m_i]: \quad \rho\frac{Dw_i}{Dt} = -\underline{\nabla}\cdot\underline{\dot{m}}''_{iD} + M_i \nu_i \dot{n}'''_{kp}$$

$$PR: \quad d\bar{h} = \bar{c}_p dT + \left(\bar{v} - T\frac{\partial \bar{v}}{\partial T}\right)dp + \sum h_i dw_i$$

$[H] - \sum h_i [m_i]$:

$$\rho\bar{c}_p\frac{DT}{Dt} + \left(1 - \frac{\partial \ln \bar{v}}{\partial \ln T}\right)\frac{Dp}{Dt} =$$

$$= -\sum \underline{\dot{m}}''_{iD}\cdot\underline{\nabla}h_i - \underline{\nabla}\cdot\underline{\dot{Q}}'' + \sum \underline{\dot{m}}''_{iD}\cdot\underline{f}_i - \underline{\underline{\tau}}:\underline{\nabla}\underline{v} + \frac{Dp}{Dt} - \sum \nu_i H_i \dot{n}'''_{kp}$$

$$\boxed{\rho\bar{c}_p\frac{DT}{Dt} = -\sum \underline{\dot{m}}''_{iD}\cdot\underline{\nabla}h_i - \underline{\nabla}\cdot\underline{\dot{Q}}'' - \sum \nu_i H_i \dot{n}'''_{kp} + \sum \underline{\dot{m}}''_{iD}\cdot\underline{f}_i - \underline{\underline{\tau}}:\underline{\nabla}\underline{v} + \frac{\partial \ln \bar{v}}{\partial \ln T}\frac{Dp}{Dt}}$$

9.3.1 Momentum Balance

We have now arrived at the column-3 part of this chapter where we shall apply the linear laws for molecular transport of momentum, matter and heat introduced as column-3 principles in cell 1.3. They will be incorporated in the momentum, mole and heat balances to yield descriptions of $\underline{v}$, c_i and T as functions of time and place.

In this section we start with the BE [m$\underline{v}$] as formulated in Sect. 9.2.1 and the linear transport law for $\underline{\underline{\tau}}$, the momentum-flux tensor due to shear in a newtonian fluid as postulated in Sect. 1.3.2.

For μ = constant and ρ = constant the net influx of momentum by shear as occurring in [m$\underline{v}$] can be rewritten as the product of the viscosity μ and the Laplacian of the velocity $\underline{v}$. The momentum balance [m$\underline{v}$] reduces to the Navier–Stokes equation.

Another special case arises when viscous effects are negligible. Then $\underline{\underline{\tau}} = 0$ and [m$\underline{v}$] reduces to the Euler equation.

$$[m\underline{v}] \quad \boxed{\rho\frac{D\underline{v}}{Dt} = -(\underline{\nabla}p + \underline{\nabla}\cdot\underline{\underline{\tau}}) + \rho\underline{g}}$$

$$\rho\frac{D\underline{v}}{Dt} = \rho\left(\frac{\partial\underline{v}}{\partial t} + \underline{v}\cdot\underline{\nabla}\underline{v}\right) = \frac{\partial}{\partial t}(\rho\underline{v}) + \underline{\nabla}\cdot(\rho\underline{v}\underline{v})$$

linear transport law:

$$\boxed{\underline{\underline{\tau}} = -\mu\{\underline{\nabla}\underline{v} + (\underline{\nabla}\underline{v})^T - \tfrac{2}{3}(\underline{\nabla}\cdot\underline{v})\underline{\underline{\delta}}\}}$$

Navier–Stokes equation

$\mu = \text{constant}, \quad \rho = \text{constant in} \; -\underline{\nabla}\cdot\underline{\underline{\tau}}$

$$[m]: \; \frac{\partial\rho}{\partial t} = -\underline{\nabla}\cdot(\rho\underline{v}), \quad \underline{\nabla}\cdot\underline{v} = 0$$

$$-\underline{\nabla}\cdot\underline{\underline{\tau}} = -\frac{\partial}{\partial x_l}\tau_{lm}\delta_m = \frac{\partial}{\partial x_l}\left\{\mu\left(\frac{\partial}{\partial x_l}v_m + \frac{\partial}{\partial x_m}v_l\right)\right\}\underline{\delta}_m$$

$$= \mu\left(\frac{\partial^2}{\partial x_l^2}v_m + \frac{\partial}{\partial x_m}\frac{\partial}{\partial x_l}v_l\right)\underline{\delta}_m$$

$$= \mu\left\{\nabla^2 v_m + \frac{\partial}{\partial x_m}(\underline{\nabla}\cdot\underline{v})\right\}\underline{\delta}_m = \mu\nabla^2\underline{v}$$

$$\boxed{\rho\frac{D\underline{v}}{Dt} = \mu\nabla^2\underline{v} - \underline{\nabla}p + \rho\underline{g}}$$

Euler equation

$$\underline{\underline{\tau}} = 0 \quad \boxed{\rho\frac{D\bar{v}}{Dt} = -\underline{\nabla}p + \rho\underline{g}}$$

A number of simplifying conditions consecutively introduced, lead to the Bernouilli equation.
When $\underline{v}$ is time-independent at a stationary point and $\underline{\underline{\tau}} = 0$ (no viscous effects), the momentum balance can be rewritten as the first simplified form.
The vector $\underline{v} . \underline{\nabla}\underline{v}$, the change of $\underline{v}$ with time when following the fluid motion in a continuous velocity field, can be rearranged using the definitions of the vector product $\underline{u} \times \underline{v}$ of two vectors and the rotation $\underline{\nabla} \times \underline{v}$ of a vector (see appendix on vectors and tensors). The summations in the definitions are over the permutations 1,2,3, 2,3,1 and 3,1,2 of the coordinate indices. The component of $\underline{\nabla} \times \underline{v}$ in the direction of $\underline{\delta}_1$ can be interpreted as the contour integral of $\underline{v} . \mathrm{d}(\mathrm{d}\underline{x})$ per unit area enclosed, for an infinitesimal surface in a plane perpendicular to $\underline{\delta}_1$. The direction of rotation is the same as for a right-handed screw moving in the direction of the unit vector $\underline{\delta}_1$.
When rearranging the component of $\underline{v} . \underline{\nabla}\underline{v}$ in the direction of $\underline{\delta}_1$, $v^2 = v_1^2 + v_2^2 + v_3^2$ and components of $\underline{\nabla} \times \underline{v}$ appear. The relation between the components of three vectors in the same direction leads to the final identity involving three vectors.
When the flow is free of rotation or $\underline{\nabla} \times \underline{v} = 0$, the change of $\underline{v}$ with time when following the fluid motion in a continuous velocity field, reduces to the gradient of the kinetic energy. When in addition ρ = constant the Bernouilli equation emerges. In Sect. 7.2.4 this equation was already found for flow in one direction.

Bernoulli equation

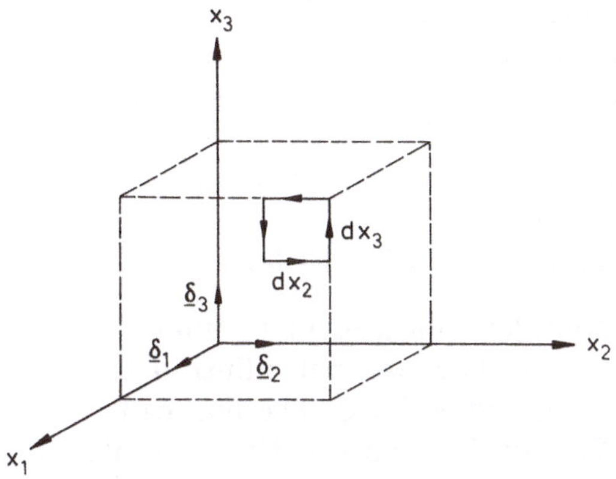

$\underline{v}$ is time-independent $\quad \rho\underline{v}\,.\,\underline{\nabla}\underline{v} = -\,\underline{\nabla}p - \rho\underline{\nabla}(gh)$

$\underline{\underline{\tau}} = 0$

vector product: $\quad \underline{u} \times \underline{v} = \sum \underline{\delta}_1(u_2 v_3 - u_3 v_2)$

rotation of a vector: $\quad \underline{\nabla} \times \underline{v} = \sum \underline{\delta}_1 \left(\frac{\partial}{\partial x_2} v_3 - \frac{\partial}{\partial x_3} v_2 \right)$

change of $\underline{v}$ with time when following the fluid motion:

$$\underline{v}\,.\,\underline{\nabla}v_1 = \left(v_1 \frac{\partial}{\partial x_1} + v_2 \frac{\partial}{\partial x_2} + v_3 \frac{\partial}{\partial x_3} \right) v_1$$

$$= \frac{1}{2}\frac{\partial}{\partial x_1}(v_1^2 + v_2^2 + v_3^2) - \left\{ v_2 \left(\frac{\partial}{\partial x_1} v_2 - \frac{\partial}{\partial x_2} v_1 \right) - v_3 \left(\frac{\partial}{\partial x_3} v_1 - \frac{\partial}{\partial x_1} v_3 \right) \right\}$$

$$= \frac{\partial}{\partial x_1}(\tfrac{1}{2} v^2) - \{ v_2(\underline{\nabla} \times \underline{v})_3 - v_3(\underline{\nabla} \times \underline{v})_2 \}$$

$$= \frac{\partial}{\partial x_1}(\tfrac{1}{2} v^2) - \{ \underline{v} \times (\underline{\nabla} \times \underline{v}) \}_1$$

$$\underline{\underline{v}\,.\,\underline{\nabla}\underline{v} = \underline{\nabla}(\tfrac{1}{2} v^2) - \underline{v} \times (\underline{\nabla} \times \underline{v})}$$

$\underline{\nabla} \times \underline{v} = 0 \quad\quad \rho\underline{\nabla}(\tfrac{1}{2} v^2) = -\,\underline{\nabla}p - \rho\underline{\nabla}(gh)$

$\underline{v}$ = time-independent $\quad$ $\boxed{p + \tfrac{1}{2}\rho v^2 + \rho g h = \text{constant}}$

$\underline{\underline{\tau}} = 0$

$\underline{\nabla} \times \underline{v} = 0$

ρ = constant

9.3.2 Diffusion

Before we turn to the mole balance of a component, we shall explore the linear law for transport of matter as postulated in Sect. 1.3.3 and the expression for $\underline{\nabla} A_i$, the gradient of the diffusion potential of component i, as derived in Sect. 9.2.3. There are N transport equations for a mixture of N components of which N − 1 equations are independent ($\Sigma x_i \underline{\nabla} A_i = 0$ and $\Sigma x_i \alpha_i = 0$).
We shall insert column-3 PR's for μ_i.

First we consider diffusion in a binary system. Consistent with the findings in Sect. 9.2.3, the net diffusion flux of component 1, $\underline{\dot{n}}''_{1D} = 0$ when $\underline{\nabla} A_1 = 0$ and $\underline{\nabla} T = 0$. The flux can be seen to be driven by the gradients $\underline{\nabla} x_1$, $\underline{\nabla} p$, $\underline{\nabla} \phi$ and $\underline{\nabla} T$. For $V_1 = V_2 = \bar{V}$ and a centrifugal field with a constant angular velocity ω, the contribution by pressure diffusion can be further rearranged.

In many engineering applications pressure diffusion, forced diffusion in an electrical field and thermo-diffusion do not play a role. Further the remaining concentration diffusion can often be reduced to Fick's law with diffusivity $\mathbb{D}_i$. This applies to a binary mixture, a mixture in which only dilute components are diffusing, and to an ideal mixture with constant Stefan–Maxwell diffusivities D_{ij}.

linear transport law for i

$$\boxed{\sum_j \frac{x_j \underline{\dot{n}}_i'' - x_i \underline{\dot{n}}_j''}{cD_{ij}} = -x_i \frac{\underline{\nabla} A_i}{RT} - \alpha_i x_i \frac{\underline{\nabla} T}{T}}$$

$$\underline{\dot{n}}_i'' = x_i c(\underline{v}_i - \underline{v}_{ref})$$

$$\underline{\nabla} A_i = (\underline{\nabla}\mu_i)_{T,p} + M_i(v_i - \bar{v})\underline{\nabla} p + z_i \mathscr{F} \underline{\nabla}\phi$$

PR's

L: $(d\mu_i)_{T,p} = RT d\ln(\gamma_i x_i)$

G: $(d\mu_i)_{T,p} = RT d\ln(\varphi_i y_i)$

Binary system

$$\frac{x_2 \underline{\dot{n}}_{1D}'' - x_1 \underline{\dot{n}}_{2D}''}{cD_{12}} = \frac{\underline{\dot{n}}_{1D}''}{cD_{12}} = -x_1 \frac{\underline{\nabla} A_1}{RT} - x_1 \alpha_1 \frac{\underline{\nabla} T}{T}$$

$$\boxed{\underline{\dot{n}}_{1D}'' = -cD_{12}\left\{\left(1 + \frac{\partial \ln \gamma_1}{\partial \ln x_1}\right)\underline{\nabla} x_1 + x_1 M_1 (v_1 - \bar{v})\frac{\underline{\nabla} p}{RT} + x_1 z_1 \mathscr{F} \frac{\underline{\nabla}\phi}{RT} + x_1 \alpha_1 \frac{\underline{\nabla} T}{T}\right\}}$$

pressure diffusion: $V_1 = V_2 = \bar{V}$, $\underline{\nabla} p = \rho\omega^2 \underline{r}$

$$\underline{\dot{n}}_{1D}'' = -\frac{cD_{12}}{RT} x_1 M_1 (v_1 - \bar{v})\underline{\nabla} p = -\frac{cD_{12}}{RT} x_1 x_2 \frac{M_1 M_2}{\bar{M}}(v_1 - v_2)\underline{\nabla} p$$

$$= -\frac{cD_{12}}{RT} x_1 x_2 (M_2 - M_1)\omega^2 \underline{r}$$

Fick's law $\boxed{\underline{\dot{n}}_{iD}'' = -c\mathbb{D}_i \underline{\nabla} x_i}$

ideal binary mixture $\boxed{\mathbb{D}_1 = D_{12}}$

$x_i \ll x_j$, $\underline{v}_j = \underline{v}$: $\sum_j \frac{x_j \underline{\dot{n}}_{iD}''}{cD_{ij}} = -\underline{\nabla} x_i$ $\boxed{\mathbb{D}_i = \left(\sum_j \frac{x_j}{D_{ij}}\right)^{-1}}$

equal D_{ij}'s, $\frac{\underline{\dot{n}}_{iD}''}{cD} = -\underline{\nabla} x_i$ $\boxed{\mathbb{D}_i = D}$

$\gamma_i = 1$ or $\varphi_i = 1$

Forced diffusion in an electrical field is illustrated by the diffusion of Zn^{2+} ions near a Zn-electrode in an aqueous $ZnCl_2$ solution. H_2O molecules and Cl^- ions have zero transport fluxes, while locally electroneutrality is maintained.

The flux $\underline{\dot{n}}_i''$ of a component i through stagnant components j is opposite to $\underline{\nabla}A_i$ and has the same sign as each $\underline{\nabla}A_j$.
Applying the transport equations to Cl^- and H_2O in an ideal solution, we can couple the flux of Zn^{2+} with the gradient $\underline{\nabla}x_1$ of its mole fraction, and the gradient $\underline{\nabla}\phi$ with $\underline{\nabla}(\ln x_1)$.

Forced diffusion in an electrical field

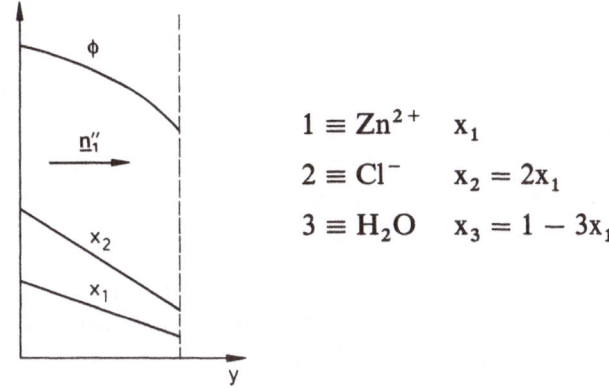

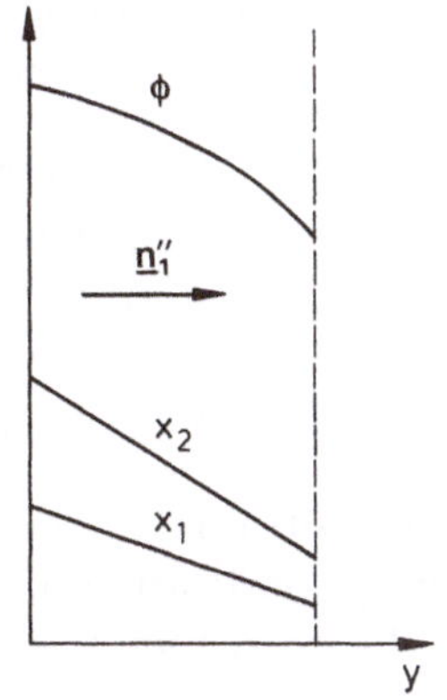

$1 \equiv Zn^{2+} \quad x_1$

$2 \equiv Cl^- \quad x_2 = 2x_1$

$3 \equiv H_2O \quad x_3 = 1 - 3x_1$

diffusing component i: Zn^{2+}

$$\sum_{j \neq i} \frac{x_j \dot{\underline{n}}_i'' - x_i \dot{\underline{n}}_j''}{cD_{ij}} = -x_i \frac{\underline{\nabla} A_i}{RT}, \quad \sum_{j \neq i} \frac{x_j}{cD_{ij}} \dot{\underline{n}}_i'' = -x_i \frac{\underline{\nabla} A_i}{RT}$$

$$\boxed{\dot{\underline{n}}_i'' = -c \left(\sum_{j \neq i} \frac{x_j}{D_{ij}} \right)^{-1} \underline{\nabla} A_i}$$

stagnant components j: Cl^-, H_2O:

$$\sum_i \frac{x_i \dot{\underline{n}}_j'' - x_j \dot{\underline{n}}_i''}{cD_{ij}} = -x_j \frac{\underline{\nabla} A_j}{RT}, \quad \frac{-x_j \dot{\underline{n}}_i''}{cD_{ij}} = -x_j \frac{\underline{\nabla} A_j}{RT}$$

$$\boxed{\dot{\underline{n}}_i'' = cD_{ij} \frac{\underline{\nabla} A_j}{RT}}$$

$j = 2, 3$:

$$\dot{\underline{n}}_1'' = cD_{12} \left\{ \underline{\nabla}(\ln x_2) - \frac{\mathscr{F} \underline{\nabla} \phi}{RT} \right\} = cD_{13} \underline{\nabla}(\ln x_3):$$

$$\boxed{\dot{\underline{n}}_1'' = -\frac{3cD_{13}}{x_3} \underline{\nabla} x_1}$$

$$\underline{\nabla}(\ln x_1) - \frac{\mathscr{F} \underline{\nabla} \phi}{RT} = -\frac{3D_{13} \underline{\nabla} x_1}{D_{12} x_3} = -\frac{3x_1 D_{13}}{x_3 D_{12}} \underline{\nabla}(\ln x_1)$$

$$\boxed{\underline{\nabla} \phi = \frac{RT}{\mathscr{F}} \left(1 + \frac{3x_1 D_{13}}{x_3 D_{12}} \right) \underline{\nabla}(\ln x_1) \approx \frac{RT}{\mathscr{F}} \underline{\nabla}(\ln x_1)}$$

9.3.3 Mole Balance

Having explored the linear law for transport of matter we shall now insert Fick's law in the mole balance for a dilute component to arrive at a description of c_i as a function of time and place.

The BE [n_i] is exact. It follows from [E] as presented in Sect. 9.1.1 by substituting $n_i/m = x_i/\bar{M}$ for $\bar{e}$. In [n_j] we neglect any production of n_j by the chemical reaction in which dilute components participate. The BE [n] is obtained by summation of [n_j] over the bulk components j while neglecting the contributions of the dilute components.

When in addition to $x_i \ll x_j$ equilibrium for the bulk components j is introduced as a second simplifying condition, the concentration diffusion flux for a dilute component i reduces to Fick's law (Sect. 9.3.2). Substitution of $\bar{M}$ = constant and Fick's law in [n_i] yields an intermediate form of [n_i].

When further c = constant and $\mathbb{D}_i$ = constant, [n_i] reduces to the final form which describes c_i as a function of time and place.

Mole balances ($x_i \ll x_j$)

$$[n_i]: \qquad \rho \frac{D}{Dt}\left(\frac{x_i}{\bar{M}}\right) = -\underline{\nabla}.\{x_i c(\underline{v}_i - \underline{v})\} + \nu_i \dot{n}'''_{kp}$$

$$[n_j]: \qquad \rho \frac{D}{Dt}\left(\frac{x_j}{\bar{M}}\right) \approx -\underline{\nabla}.\{x_j c(\underline{v}_j - \underline{v})\}$$

$$[n] \approx \sum_j [n_j]: \quad \rho \frac{D}{Dt}\left(\frac{1}{\bar{M}}\right) \approx -\underline{\nabla}.\{c(\underline{v}^* - \underline{v})\}$$

Equilibrium for bulk components j

$$\underline{v}_j \approx \underline{v} \approx \underline{v}^*: \quad x_j \approx \text{constant}, \quad \bar{M} \approx \text{constant}$$

Fick's law ($x_i \ll x_j$, $\underline{v}_j = \underline{v}$)

$$\underline{\dot{n}}''_{iD} = x_i c(\underline{v}_i - \underline{v}) = -c\mathbb{D}_i \underline{\nabla} x_i \quad \text{with } \mathbb{D}_i = \left(\sum_j \frac{x_j}{D_{ij}}\right)^{-1}$$

Mole balance for dilute component under simplifying conditions

$x_i \ll x_j$, equilibrium for bulk components

$$[n_i]: \quad \boxed{c\frac{Dx_i}{Dt} = \underline{\nabla}.(c\mathbb{D}_i \underline{\nabla} x_i) + \nu_i \dot{n}'''_{kp}}$$

$c = \text{constant}$ ($\rho = \text{constant}$), $\mathbb{D}_i = \text{constant}$

$$[n_i]: \quad \boxed{\frac{Dc_i}{Dt} = \mathbb{D}_i \nabla^2 c_i + \nu_i \dot{n}'''_{kp}}$$

9.3.4 Heat Balance

Having simplified the rigorous BE's and linear transport laws for momentum and matter, we shall do the same for heat.

We start from [Heat] as derived in Sect. 9.2.5 and the linear law for molecular transport of heat law giving the heat flux $\underline{\dot{Q}}''$ as postulated in Sect. 1.3.3.

As already noted in Sect. 9.2.5 the terms in [Heat] which involve the diffusion flux $\underline{\dot{m}}''_{iD}$ are usually negligible. In many applications the temperature rise by reversible compression can be ignored as well. Further the Dufour or diffusion-thermo effect plays only a role in special cases.

For zero diffusion and compression effects, the heat balance reduces to a form which describes the rate of temperature increase of an element following the fluid motion by heat conduction, release of heat by a chemical reaction and dissipation of mechanical energy.

A further simplification is obtained for λ = constant.

[heat]:

$$\rho\bar{c}_p \frac{DT}{Dt} = -\sum \underline{\dot{m}}''_{iD} \cdot \underline{\nabla} h_i - \underline{\nabla} \cdot \underline{\dot{Q}}'' - \sum \nu_i H_i \dot{n}'''_{kp} + \sum \underline{\dot{m}}''_{iD} \cdot \underline{f}_i - \underline{\underline{\tau}} : \underline{\nabla}\underline{v} + \frac{\partial \ln \bar{v}}{\partial \ln T} \frac{Dp}{Dt}$$

linear transport law

$$\underline{\dot{Q}}'' = -\lambda \underline{\nabla} T + RT \sum \alpha_i \underline{\dot{n}}''_i$$

zero diffusion and compression effects

$$\boxed{\rho\bar{c}_p \frac{DT}{Dt} = \underline{\nabla} \cdot (\lambda \underline{\nabla} T) - \sum \nu_i H_i \dot{n}'''_{kp} - \underline{\underline{\tau}} : \underline{\nabla}\underline{v}}$$

λ is constant

$$\boxed{\rho\bar{c}_p \frac{DT}{Dt} = \lambda \nabla^2 T - \sum \nu_i H_i \dot{n}'''_{kp} - \underline{\underline{\tau}} : \underline{\nabla} v}$$

9.3.5 Set of Simplified Transport Equations

In this section the set of simplified transport equations is presented. It consists of BE's, equations for molecular transport and definitions of transfer coefficients for momentum, matter and heat.
The momentum, mole and heat balances are further simplifications of the equations presented previously in this cell. Only terms for accumulation, convection and molecular transport have remained. The equations describe $\underline{v}$, c_i and T as functions of time and place.
The simplified equations for molecular transport were encountered before. They predict fluxes which are down the gradients of $\underline{v}$, c_i and T. The set of simplified transport equations is completed by the definitions of the Fanning friction factor f, the mass transfer coefficient k_i for component i and the heat transfer coefficient α. They are formal coefficients to give transversal fluxes from a wall or interface to the bulk of a phase at a given potential difference between interface and bulk. Right at an interface transversal transport is usually only molecular which gives rise to the boundary conditions indicated.

momentum	$[m\underline{v}]$:	$$\boxed{\rho\frac{\partial \underline{v}}{\partial t} = -\rho\underline{v}\cdot\underline{\nabla}\underline{v} + \mu\nabla^2\underline{v}}$$
	Newton's law:	$\underline{\underline{\tau}} = -\mu\{\underline{\nabla}\underline{v} + (\underline{\nabla}\underline{v})^T\}$
	transfer:	$\tau_n = -\mu\frac{\partial v_l}{\partial x_n} = \frac{f}{2}\rho v_\infty(v_w - v_\infty)$
matter	$[n_i]$:	$$\boxed{\frac{\partial c_i}{\partial t} = -\underline{v}\cdot\underline{\nabla}c_i + \mathbb{D}_i\nabla^2 c_i}$$
	Fick's law:	$\underline{\dot{n}}''_{iD} = -\mathbb{D}_i\underline{\nabla}c_i$
	transfer:	$\dot{n}''_{iD,n} = -\mathbb{D}_i\frac{\partial c_i}{\partial x_n} = k_i(c_{i,w} - c_{i,\infty})$
heat	[heat]:	$$\boxed{\rho\bar{c}_p\frac{\partial T}{\partial t} = -\rho\bar{c}_p\underline{v}\cdot\underline{\nabla}T + \lambda\nabla^2 T}$$
	Fourier's law:	$\underline{\dot{Q}}'' = -\lambda\underline{\nabla}T$
	transfer:	$\dot{Q}''_n = -\lambda\frac{\partial T}{\partial x_n} = \alpha(T_w - T_\infty)$

We shall apply the set of simplified transport equations to two examples to illustrate that from the inherent analogy between the equations for momentum, matter and heat conclusions can be drawn without explicitly solving the equations.

In the first example we describe the approach to a new equilibrium of a system subjected to a step change in longitudinal velocity v_1, component concentration c_i or temperature T at its interface.
Accumulation and molecular transport play a role, convective transport does not. The momentum, mole and heat balances can be simplified accordingly. The kinematic viscosity ν and the thermal diffusivity a emerge as analogues of the diffusivity $\mathbb{D}_i$.
Each of the three cases can be described by the same generalised partial differential equation when introducing the generalised diffusivity $\mathcal{D}$ which stands for ν, $\mathbb{D}_i$ or a.
The partial differential equation can subsequently be made dimensionless for geometrically similar cases when we introduce the dimensionless time $t' = \mathcal{D}t/L^2$. It follows that for each case the relative departure from equilibrium is the same function of t′ and the dimensionless coordinates. Dependent on the choice of the length scale L and the geometry of the system, the remaining maximum departure from equilibrium in the system is less than 10 to 1% for a Fourier number $t' = 1$.

Equations for zero convection

[m$\underline{v}$]: $$\frac{\partial v_1}{\partial t} = \nu\nabla^2 v_1, \quad \nu = \frac{\mu}{\rho}$$

[n_i]: $$\frac{\partial c_i}{\partial t} = \mathbb{D}_i \nabla^2 c_i$$

[heat]: $$\frac{\partial T}{\partial t} = a\nabla^2 T, \quad a = \frac{\lambda}{\rho \bar{c}_p}$$

generalised partial differential equation:

$$\frac{\partial \check{E}}{\partial t} = \mathscr{D}\nabla^2 \check{E}$$

dimensionless partial differential equation:

$$\frac{\partial \check{E}'}{\partial t} = \frac{\mathscr{D}}{L^2}(\nabla')^2 \check{E}'$$

$$\frac{\partial \check{E}'}{\partial t'} = (\nabla')^2 \check{E}'$$

dimensionless time $$\boxed{t' = \frac{\mathscr{D}t}{L^2}}$$

In the second example we consider time-independent fields for $\underline{v}$, c_i and T near an interface and transversal interface transport of momentum, matter and heat. In the BE's the accumulation terms vanish.
When we convert the BE's and boundary conditions to dimensionless forms, various dimensionless groups appear as coefficients in the equations. For momentum this is Reynolds which represents the ratio of convective to molecular transport of momentum. For matter the Schmidt number is a property which reflects the ratio of molecular transport of momentum to molecular transport of matter, while the Sherwood number compares interface transfer of matter with molecular transport of matter. Analogously the Prandtl and Nusselt numbers present themselves for heat.
Inspection of the dimensionless equations reveals that for geometrically similar cases the dimensionless velocity field is entirely fixed by the value of Re and that the Fanning friction factor f is a function of Re.
When in addition Sc is given, the dimensionless concentration field is determined as well. Consequently Sh is a function of Re and Sc, usually of the form $Sh = cRe^m Sc^n$, where c and the exponents are constants for a given geometry.
Finally, it follows from the analogy between transport of heat and matter that Nu is the same function of Re and Pr as Sh is of Re and Sc.

Dimensionless equations for zero accumulation

momentum	$0 = -\mathrm{Re}\,\underline{v}'.\underline{\nabla}'\underline{v}' + (\nabla')^2\underline{v}'$ $-\frac{\partial v_i'}{\partial x_n'} = \frac{f}{2}\mathrm{Re}(v_w' - v_\infty')$	$\mathrm{Re} = \frac{v_\infty L}{\nu}$
matter	$0 = -\mathrm{Re}\,\mathrm{Sc}\,\underline{v}'.\underline{\nabla}'c_i' + (\nabla')^2 c_i'$ $-\frac{\partial c_i'}{\partial x_n'} = \mathrm{Sh}(c_{i,w}' - c_{i,\infty}')$	$\mathrm{Sc} = \frac{\nu}{\mathbb{D}_i}$ $\mathrm{Sh} = \frac{k_i L}{\mathbb{D}_i}$
heat	$0 = -\mathrm{Re}\,\mathrm{Pr}\,\underline{v}'.\underline{\nabla}'T' + (\underline{\nabla}')^2 T'$ $-\frac{\partial T'}{\partial x_n'} = \mathrm{Nu}(T_w' - T_\infty')$	$\mathrm{Pr} = \frac{\nu}{a}$ $\mathrm{Nu} = \frac{\alpha L}{\lambda}$

A1 Vectors

A vector is a quantity with a magnitude and a direction. Examples of vectors encountered are an infinitesimal length and area, $d\underline{z}$ and $d\underline{A}$, component, mass average and mole average velocities $\underline{v}_i$, $\underline{v}$ and $\underline{v}^*$, a differential force $d\underline{F}$, the acceleration $\underline{g}$ due to gravity, and fluxes such as the generalised fluxes of an extensive $\dot{\underline{E}}''_x$, $\dot{\underline{E}}''$ and $\dot{\underline{E}}''_{im}$, the work flux $\dot{\underline{W}}''$, the flux of momentum $\underline{\tau}_1$ in the 1-direction and the fluxes of matter and heat, $\dot{\underline{n}}''_i$ and $\dot{\underline{Q}}''$.

In general a vector $\underline{a}$ can be represented by a directed line segment or an arrow. The magnitude $|\underline{a}| = a$ corresponds to the length of the segment. The unit vector $\underline{a}/a$ is of unit length. The negative $-\underline{a}$ of a vector $\underline{a}$ has the same length but points in the opposite direction.

Vector addition and subtraction are defined by parallelogram or equivalent triangle constructions. When $\underline{b}$ is placed where $\underline{a}$ ends, the end point of the vector sum $\underline{a} + \underline{b}$ is found (a vector may be shifted provided that its length and direction are preserved). The vector difference $\underline{a} - \underline{b}$ is what has to be added to $\underline{b}$ to obtain $\underline{a}$. It follows that vector addition is commutative and associative: $\underline{a} + \underline{b} = \underline{b} + \underline{a}$ and $\underline{a} + (\underline{b} + \underline{c}) = (\underline{a} + \underline{b}) + \underline{c}$.

The scalar product $\underline{a}.\underline{b}$ of two vectors is the length of one of them multiplied by the scalar projection of the other upon it. It contains the cosine of φ, the angle between the positive directions of $\underline{a}$ and $\underline{b}$. From its definition it can be concluded that scalar multiplication is commutative and distributive over addition: $\underline{a}.\underline{b} = \underline{b}.\underline{a}$ and $\underline{a}.(\underline{b} + \underline{c}) = \underline{a}.\underline{b} + \underline{a}.\underline{c}$. Scalar and vector projections of $\underline{b}$ on $\underline{a}$ can be written in terms of the scalar product of $\underline{b}$ and the unit vector $\underline{a}/a$.

Magnitude, unit vector, negative of a vector

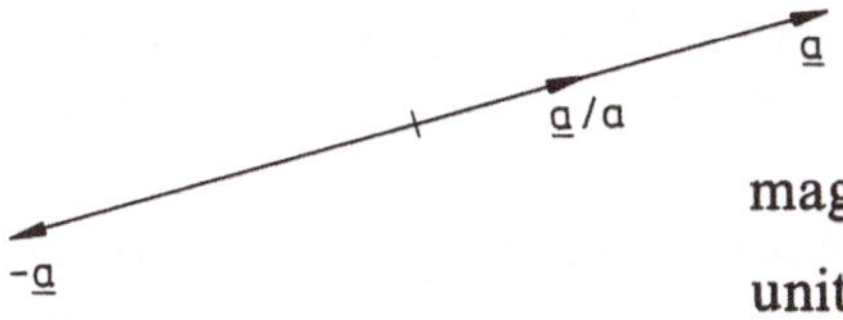

magnitude: $|\underline{a}| = a$

unit vector in direction of $\underline{a}$: $\underline{a}/a$

negative of $\underline{a}$: $-\underline{a} = -a(\underline{a}/a)$

Addition and subtraction $\underline{a} + \underline{b}$, $\underline{a} - \underline{b}$

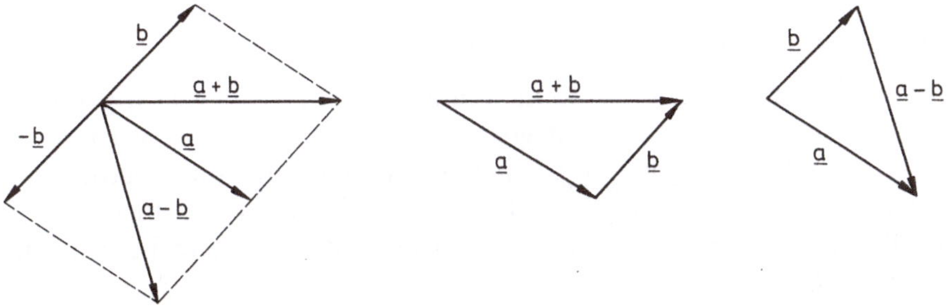

Scalar product $\underline{a} . \underline{b}$

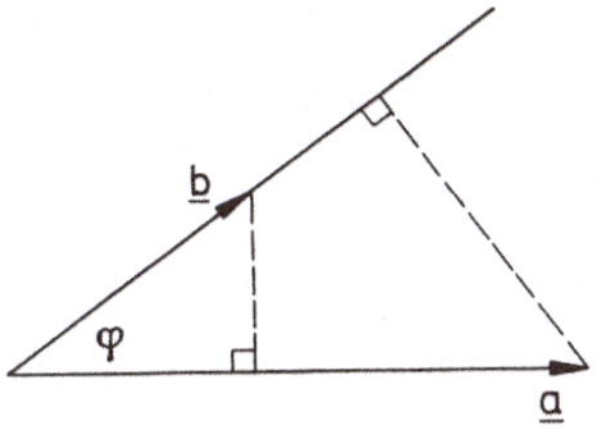

$$\boxed{\underline{a} . \underline{b} = ab \cos \varphi}$$

scalar projection of $\underline{b}$ on $\underline{a}$:

$$b \cos \varphi = \underline{b} . (\underline{a}/a)$$

vector projection of $\underline{b}$ on $\underline{a}$:

$$b \cos \varphi (\underline{a}/a) = \underline{b} . (\underline{a}/a) (\underline{a}/a)$$

$$\underline{a} . (\underline{b} + \underline{c}) = \underline{a} . \underline{b} + \underline{a} . \underline{c}$$

Another product is the vector product $\underline{a} \times \underline{b}$. In defining this product we shall apply rotation of a vector in a plane. The direction of rotation is defined by the unit vector $\underline{\delta}_n$ normal to that plane and pointing in the direction that a right-handed screw would advance when turned as the vector in the plane.
The vector product $\underline{a} \times \underline{b}$ is obtained by projecting $\underline{b}$ on a plane perpendicular to $\underline{a}$, rotating the projected vector over 90° about $\underline{a}$ and in the direction of $\underline{a}$ and, finally, multiplying the rotated vector by a. The vector $\underline{a} \times \underline{b}$ has as magnitude the area of parallelogram formed by $\underline{a}$ and $\underline{b}$ and points in the direction of the rotation from $\underline{a}$ to $\underline{b}$ by the shortest route. It follows that vector multiplication is distributive over addition and anticommutative:

$$\underline{a} \times (\underline{b} + \underline{c}) = \underline{a} \times \underline{b} + \underline{a} \times \underline{c} \quad \text{and} \quad \underline{b} \times \underline{a} = -\underline{a} \times \underline{b}.$$

For the unit vector $\underline{\delta}_n$ normal to the plane of $\underline{a}$ and $\underline{b}$, the vector product can be written as $\underline{a} \times \underline{b} = ab \sin\varphi \, \underline{\delta}_n$, where φ is the angle of rotation from $\underline{a}$ to $\underline{b}$ in the direction of $\underline{\delta}_n$. When we introduce the unit vector $\underline{\delta}_1$ in the plane of $\underline{a}$ and $\underline{b}$, $\underline{a} \times \underline{b}$ can alternatively be expressed in terms of α and β, the angles of rotation from $\underline{\delta}_1$ to $\underline{a}$ and from $\underline{\delta}_1$ to be $\underline{b}$ in the direction of $\underline{\delta}_n$.

When the vectors $\underline{a}$ and $\underline{b}$ in a plane with unit normal vector $\underline{\delta}_n$ are projected on another plane with unit normal vector $\underline{\delta}'_n$, the vector product $\underline{a}' \times \underline{b}'$ of the projected vectors is readily shown to be equal to the projection of $\underline{a} \times \underline{b}$ on $\underline{\delta}'_n$. Upon projection the area of the parallelogram formed by the two vectors has to be multiplied by $\cos\psi$, where ψ is the angle between the two planes or between $\underline{\delta}_n$ and $\underline{\delta}'_n$.

Vector product: $\underline{a} \times \underline{b}$

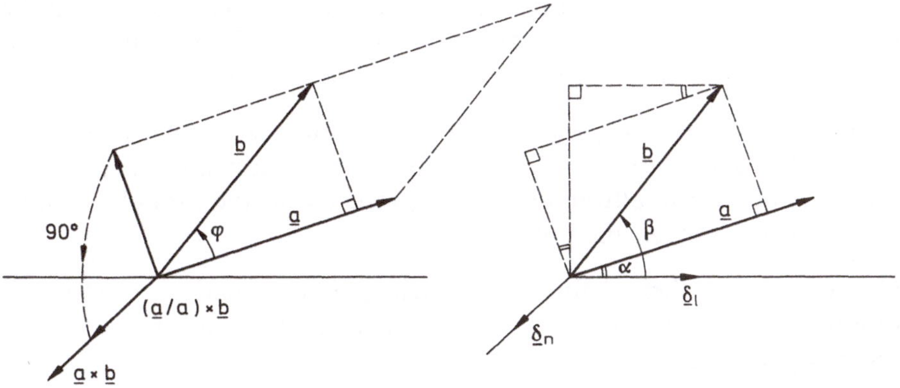

$$\underline{a} \times \underline{b} = ab \sin \varphi \, \underline{\delta}_n = a\{\cos \alpha(b \sin \beta - b \cos \beta \tan \alpha)\} \, \underline{\delta}_n$$

$$\boxed{\underline{a} \times \underline{b} = ab \sin \varphi \, \underline{\delta}_n = ab(\sin \beta \cos \alpha - \sin \alpha \cos \beta) \, \underline{\delta}_n}$$

Vector product of projections: $\underline{a}' \times \underline{b}'$

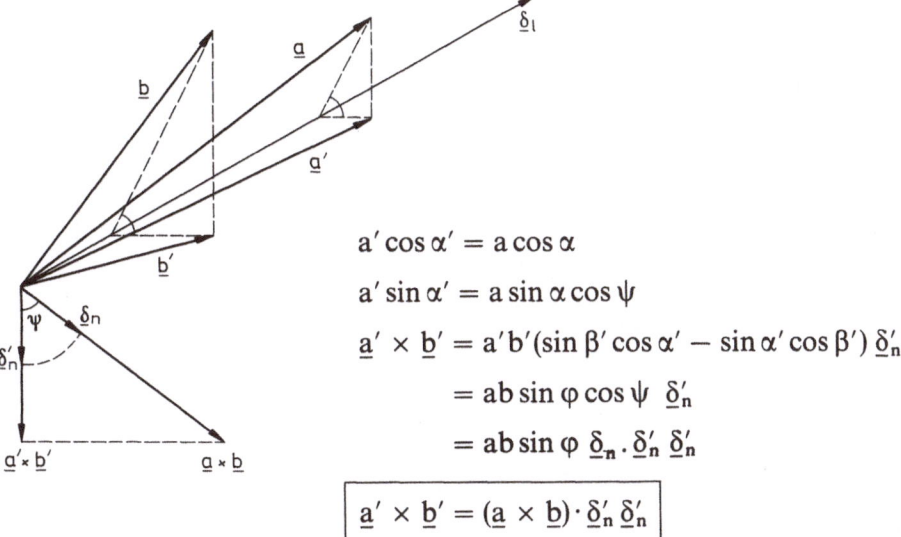

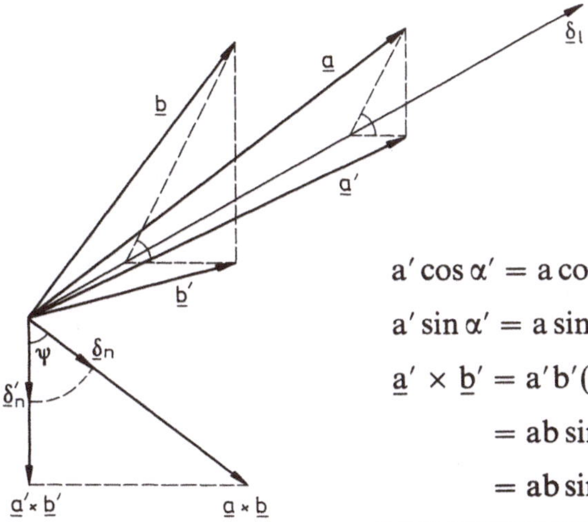

$$a' \cos \alpha' = a \cos \alpha$$

$$a' \sin \alpha' = a \sin \alpha \cos \psi$$

$$\underline{a}' \times \underline{b}' = a'b'(\sin \beta' \cos \alpha' - \sin \alpha' \cos \beta') \, \underline{\delta}'_n$$

$$= ab \sin \varphi \cos \psi \;\; \underline{\delta}'_n$$

$$= ab \sin \varphi \;\; \underline{\delta}_n . \underline{\delta}'_n \, \underline{\delta}'_n$$

$$\boxed{\underline{a}' \times \underline{b}' = (\underline{a} \times \underline{b}) \cdot \underline{\delta}'_n \, \underline{\delta}'_n}$$

$$|\underline{a}' \times \underline{b}'| = |\underline{a} \times \underline{b}| . |\cos \psi|$$

A2 Vectors Represented as Projections on Orthogonal Axes

We shall apply a system of three orthogonal axes. When defining two positive directions by the unit vectors $\underset{\sim}{\delta}_1$ and $\underset{\sim}{\delta}_2$, a choice can still be made for the third positive direction. We shall restrict ourselves to a right-handed system for which $\underset{\sim}{\delta}_3 = \underset{\sim}{\delta}_1 \times \underset{\sim}{\delta}_2$ ($\varphi = 90°$).

A vector $\underset{\sim}{a}$ is completely represented by the three scalar projections on the unit vectors or $\underset{\sim}{a} = (a_1, a_2, a_3)$. From the definition of vector addition it follows that $\underset{\sim}{a}$ can alternatively be represented as the sum of the three vector projections on the unit vectors.

Consistent with their definitions by triangle constructions, vector addition $\underset{\sim}{a} + \underset{\sim}{b}$ and subtraction $\underset{\sim}{a} - \underset{\sim}{b}$ reduce to addition and subtraction of the components.

The scalar product of two different unit vectors is zero ($\varphi = 90°$) and the product of a unit vector with itself is unity ($\varphi = 0°$). By writing $\underset{\sim}{a}$ and $\underset{\sim}{b}$ as sums of vector projections, the scalar product becomes a product of components summed over the three directions.

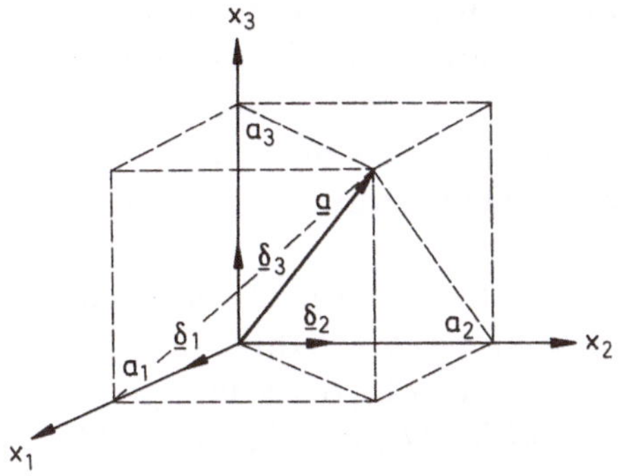

$$\underline{\delta}_3 = \underline{\delta}_1 \times \underline{\delta}_2$$

Scalar projections

$$\underline{a} . \underline{\delta}_1 = a \cos \alpha_1 = a_1$$

$$\underline{a} . \underline{\delta}_2 = a \cos \alpha_2 = a_2$$

$$\underline{a} . \underline{\delta}_3 = a \cos \alpha_3 = a_3$$

$$\boxed{a^2 = a_1^2 + a_2^2 + a_3^2}$$

Vector representations

$$\boxed{\underline{a} = (a_1, a_2, a_3) = a_1 \underline{\delta}_1 + a_2 \underline{\delta}_2 + a_3 \underline{\delta}_3}$$

unit vectors: $\underline{\delta}_1 = (1, 0, 0)$, $\underline{\delta}_2 = (0, 1, 0)$, $\underline{\delta}_3 = (0, 0, 1)$

Addition and subtraction

$$\underline{a} + \underline{b} = (a_1 + b_1, a_2 + b_2, a_3 + b_3)$$

$$\underline{a} - \underline{b} = (a_1 - b_1, a_2 - b_2, a_3 - b_3)$$

Scalar product

$\underline{\delta}_l . \underline{\delta}_m = \delta_{lm}$ (Kronecker delta)

$$\delta_{lm} = 1 \quad \text{if } l = m, \quad \delta_{lm} = 0 \quad \text{if } l \neq m$$

$$\underline{a} . \underline{b} = (a_1 \underline{\delta}_1 + a_2 \underline{\delta}_2 + a_3 \underline{\delta}_3) . (b_1 \underline{\delta}_1 + b_2 \underline{\delta}_2 + b_3 \underline{\delta}_3)$$

$$\boxed{\underline{a} . \underline{b} = a_1 b_1 + a_2 b_2 + a_3 b_3}$$

The vector product $\underline{a} \times \underline{b}$ can be expressed in the components of $\underline{a}$ and $\underline{b}$ using the identity of the product $\underline{a}' \times \underline{b}'$ of vectors projected on a plane and the projection of $\underline{a} \times \underline{b}$ on the unit vector normal to that plane. For projection on the plane of $\underline{\delta}_2$ and $\underline{\delta}_3$ this leads to $(\underline{a} \times \underline{b})_1 = a_2b_3 - a_3b_2$. The other components are found by cyclical permutation of the coordinate indices.
An alternative way to find $\underline{a} \times \underline{b}$ in terms of the components of $\underline{a}$ and $\underline{b}$ is by writing $\underline{a}$ and $\underline{b}$ as sums of vector projections on the three orthogonal unit vectors. The vector projection on $\underline{\delta}_1$, $(\underline{a} \times \underline{b})_1 \underline{\delta}_1$ is composed of contributions of $\underline{\delta}_2 \times \underline{\delta}_3 = \underline{\delta}_1 (\varphi = 90°)$ and $\underline{\delta}_3 \times \underline{\delta}_2 = -\underline{\delta}_1 (\varphi = -90°)$.

Vector product

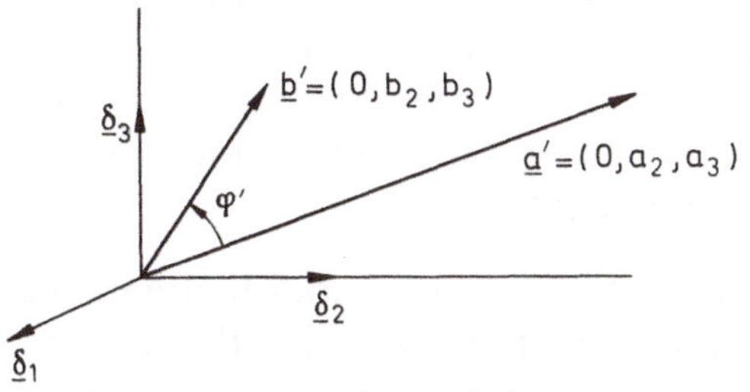

$$\underline{a}' \times \underline{b}' = a'b' \sin\varphi' \underline{\delta}_1 = a'b'(\sin\alpha' \cos\beta' - \sin\beta' \cos\alpha')\underline{\delta}_1$$

$$= (a_2 b_3 - a_3 b_2)\underline{\delta}_1$$

$$\underline{a}' \times \underline{b}' = \{(\underline{a} \times \underline{b}) . \underline{\delta}_1\}\underline{\delta}_1 = (\underline{a} \times \underline{b})_1 \underline{\delta}_1$$

$$\boxed{\begin{aligned}\underline{a} \times \underline{b} &= \underline{\delta}_1(a_2 b_3 - a_3 b_2) + \underline{\delta}_2(a_3 b_1 - a_1 b_3) + \underline{\delta}_3(a_1 b_2 - a_2 b_1)\\ &= \sum \underline{\delta}_1(a_2 b_3 - a_3 b_2)\end{aligned}}$$

$$\underline{a} \times \underline{b} = (a_1\underline{\delta}_1 + a_2\underline{\delta}_2 + a_3\underline{\delta}_3) \times (b_1\underline{\delta}_1 + b_2\underline{\delta}_2 + b_3\underline{\delta}_3)$$

$$(\underline{a} \times \underline{b})_1 \underline{\delta}_1 = (a_2\underline{\delta}_2) \times (b_3\underline{\delta}_3) + (a_3\underline{\delta}_3) \times (b_2\underline{\delta}_2) = (a_2 b_3 - a_3 b_2)\underline{\delta}_1$$

A3 Products Involving Vectors and Tensors

We found that the three orthogonal unit vectors $\underline{\delta}_l$ can be represented as a row or column of three components of which the l-th component is unity and the other components are zero. It is convenient to define in addition nine unit dyads $\underline{\delta}_l\underline{\delta}_m$ which can be represented as an array of nine components of which the component in the l-th row and the m-th column is unity, while the other components are zero.

Here we shall adhere to the convection that repeated coordinate indices imply summation over the three directions. In A2 we found that a vector $\underline{v}$ can be represented as the product $v_l\underline{\delta}_l$ summed over the three directions or as a row or column of three components. Analogously a tensor $\underline{\underline{\tau}}$ can be represented as a product of the tensor component and the unit dyad $\tau_{lm}\underline{\delta}_l\underline{\delta}_m$ summed over l and m or, alternatively, as an array of nine components τ_{lm}. An example of a tensor is the momentum-flux tensor $\underline{\underline{\tau}}$. The row vector $\underline{\tau}_l = \tau_{lm}\underline{\delta}_m$ represents the momentum flux in the direction of l, τ_{lm} the m-th component of $\underline{\tau}_l$. The transpose of $\underline{\underline{\tau}}$ has components of which row and column indices are interchanged. A tensor is symmetric when $\tau_{lm} = \tau_{ml}$. The unit tensor $\underline{\underline{\delta}}$ is a tensor of which the diagonal components are unity and the non-diagonal components are zero.

Using the representations of vectors and tensors in terms of $\underline{\delta}_l$ and $\underline{\delta}_l\underline{\delta}_m$, a number of products can be defined.
The dyadic product of two vectors is a tensor.
Three single-dot products are indicated: the scalar product of two vectors, the product of $\underline{\underline{\tau}}$ and $\underline{v}$ (the work flux $\dot{\underline{W}}''$ by shear forces when $\underline{\underline{\tau}}$ is the momentum-flux tensor and $\underline{v}$ the velocity), and the product of two tensors. In all three cases a scalar or component is obtained by one summation over the coordinate indices.
The double-dot product of two tensors involves two successive summations over the coordinate indices.

Unit vectors and dyads

3 unit vectors $\underline{\delta}_l$: $\underline{\delta}_1 = (1, 0, 0)$ etc.

9 unit dyads $\underline{\delta}_l\underline{\delta}_m$: $\underline{\delta}_1\underline{\delta}_2 = \begin{pmatrix} 0 & 1 & 0 \\ 0 & 0 & 0 \\ 0 & 0 & 0 \end{pmatrix}$ etc

Vectors and tensors

vector: $\underline{v} = v_l\underline{\delta}_l = (v_1, v_2, v_3)$

tensor: $\underline{\underline{\tau}} = \tau_{lm}\underline{\delta}_l\underline{\delta}_m = \begin{pmatrix} \tau_{11} & \tau_{12} & \tau_{13} \\ \tau_{21} & \tau_{22} & \tau_{23} \\ \tau_{31} & \tau_{32} & \tau_{33} \end{pmatrix}$

transpose of $\underline{\underline{\tau}}$: $\tau^T_{lm} = \tau_{ml}$

symmetric tensor: $\tau_{lm} = \tau_{ml}$

row vector: $\underline{\tau}_l = \tau_{lm}\underline{\delta}_m$

unit tensor: $\underline{\underline{\delta}} = \delta_{lm}\underline{\delta}_l\underline{\delta}_m = \begin{pmatrix} 1 & 0 & 0 \\ 0 & 1 & 0 \\ 0 & 0 & 1 \end{pmatrix}$

Products

dyadic product of two vectors: $\underline{v}\underline{w} = v_l v_m \underline{\delta}_l\underline{\delta}_m$

scalar product of two vectors: $\underline{v} . \underline{w} = v_l w_l$

single-dot peoduct of $\underline{\underline{\tau}}$ and $\underline{v}$: $\underline{\underline{\tau}} . \underline{v} = \tau_{lm} v_m \underline{\delta}_l$

single-dot product of two tensors: $\underline{\underline{\tau}} . \underline{\underline{\sigma}} = \tau_{lk}\sigma_{km}\underline{\delta}_l\underline{\delta}_m$

double-dot product of two tensors: $\underline{\underline{\tau}} : \underline{\underline{\sigma}} = \tau_{lm}\sigma_{ml}$

A4 Differential Operators

As differential operators we met the vector differential operator $\underline{\nabla}$ (nabla), the Laplacian operator $\underline{\nabla}^2 = \nabla^2$ and the substantial derivative D/Dt, the change with time when following the fluid motion.

With the aid of $\underline{\nabla}$ gradients and divergencies (net effluxes per unit volume) can be defined. The gradient turns a scalar field into a vector field and a vector field into a tensor field. The divergence of a vector is a scalar, that of a tensor a vector.

We encountered the Laplacian of a scalar field and a vector field as molecular transport terms in the BE's for momentum, matter and heat under simplifying conditions.

vector differential operator (nabla)	$\boxed{\underline{\nabla} = \underline{\delta}_l \frac{\partial}{\partial x_l}}$
Laplacian operator	$\boxed{\underline{\nabla}^2 = \nabla^2 = \frac{\partial^2}{\partial x_l^2} = \frac{\partial^2}{\partial x_1^2} + \frac{\partial^2}{\partial x_2^2} + \frac{\partial^2}{\partial x_3^2}}$
substantial derivative	$\boxed{\frac{D}{Dt} = \frac{\partial}{\partial t} + \frac{Dx_l}{Dt}\frac{\partial}{\partial x_l} = \frac{\partial}{\partial t} + v_l \frac{\partial}{\partial x_l} = \frac{\partial}{\partial t} + \underline{v} . \underline{\nabla}}$
gradient of a scalar:	$\underline{\nabla} T = \frac{\partial}{\partial x_l} T \underline{\delta}_l$
gradient of a vector:	$\underline{\nabla}\underline{v} = \frac{\partial}{\partial x_l} v_m \underline{\delta}_l \underline{\delta}_m$
divergence of a vector:	$\underline{\nabla} . \underline{\dot{E}}''_x = \frac{\partial}{\partial x_l} \dot{E}''_{x,l}$
divergence of a tensor:	$\underline{\nabla} . \underline{\underline{\tau}} = \frac{\partial}{\partial x_l} \tau_{lm} \underline{\delta}_m$
Laplacian of a scalar:	$\nabla^2 T = \frac{\partial^2}{\partial x_l^2} T = \left(\frac{\partial^2}{\partial x_1^2} + \frac{\partial^2}{\partial x_2^2} + \frac{\partial^2}{\partial x_3^2}\right) T$
Laplacian of a vector:	$\nabla^2 \underline{v} = \frac{\partial^2}{\partial x_l^2} \underline{v} = \left(\frac{\partial^2}{\partial x_1^2} + \frac{\partial^2}{\partial x_2^2} + \frac{\partial^2}{\partial x_3^2}\right) \underline{v}$
rotation of a vector:	$\underline{\nabla} \times \underline{v} = \sum \underline{\delta}_1 \left(\frac{\partial}{\partial x_2} v_3 - \frac{\partial}{\partial x_3} v_2\right)$

Structure and Contents

THEORY OF MACROSCOPIC SYSTEMS	Column 1 MOLE NUMBERS, MASSES AND GENERALISED EXTENSIVES	Column 2 FUNDAMENTAL EXTENSIVES	Column 3 PHENOMENOLOGICAL LAWS
1. Principles	1. Properties and BE's of a system 2. Equilibrium state and PR's of a system 3. Intensives associated with single-phase extensives 4. Reduced properties associated with multi-phase extensives 5. Transport of extensive E by matter	1. Immaterial transport of fundamental extensives 2. Energy transport by work 3. Production of fundamental extensives. Energy functions	1. Asymptotic phase behaviour 2. Transport of momentum 3. Transport of matter and heat
2. First Deductions	1. BE's over a time interval and per unit time	1. FBE's over a time interval 2. FBE's per unit time 3. Exergy 4. Interface transport and equilibrium conditions 5. FPR of matter 6. Chemical equilibrium condition 7. Electrochemical equilibrium condition	
3. PR's of Single-Phase Systems	1. Extensive state 2. Intensive state	1. FPR and equivalents. Reversible Heating. Functions of T, V, n_i and T, p, n_i 2. Closed system 3. Set of single-phase PR's	1. Standard properties of heating 2. μ_i-models 3. Properties of compression and mixing 4. Exergy 5. Closed ideal-gas system 6. Residual properties 7. Activity coefficients
4. PR's of Multi-Phase Systems	1. Closed system with $f_{int} = f_{ext} = 1$	1. FPR and equivalents. Degrees of freedom. Gibbs phase rule. 2. Closed system with $f_{int} = f_{ext} = 1$ 3. Two-phase binary system	1. Single-component systems 2. Binary systems 3. G—S systems 4. L—S systems 5. G—L systems 6. L—L systems
5. PR's of Reaction Systems	1. Closed single-reaction system	1. Closed-system FPR. Closed system with $f_{int} = f_{ext} = 1$. Gibbs phase rule	1. Chemical equilibrium constants 2. Electrochemical systems
6. BE's of Non-Flow Systems		1. BE's over a time interval. Charging/discharging 2. Cycles 3. Carnot cycle 4. Heating/compression cycles 5. Reversible interface transport 6. Closed-system BE's per unit time	1. Heating/cooling and compression/expansion of a closed system 2. Irreversible reaction in a closed system 3. Otto and Diesel cycles 4. Brayton and Rankine cycles 5. Stirling and Ericsson cycles
7. BE's of Continuous Plug Flow Systems		1. Description as a place-dependent cycle 2. Description as a time-dependent cycle 3. Heating/compression cycles 4. BE's per unit time	1. Brayton cycles
8. BE's of Continuous Mixed Flow Systems		1. BE's over a time interval 2. Adiabatic and isothermal mixing/seperation 3. Adiabatic and isothermal reactions 4. Reversible counterparts of a reaction in a non-flow system	1. BE's per unit time
9. BE's of Infinitesimal Systems	1. Generalised BE'S	1. FBE's 2. Mechanical energy and enthalpy balances 3. Entropy production 4. Enquilibrium concentration distributions 5. Heat balance	1. Momentum balance 2. Diffusion 3. Mole balance 4. Heat balance 5. Set of simplified transport equations